WILD ATOM

NUCLEAR TERRORISM

WILD ATOM

NUCLEAR

TERRORISM

Global Organized Crime Project

PROJECT CHAIR: WILLIAM H. WEBSTER
PROJECT DIRECTOR: ARNAUD DE BORCHGRAVE
PROJECT CODIRECTORS: ROBERT H. KUPPERMAN
                                 & ERIK R. PETERSON
PROJECT DEPUTY DIRECTOR: FRANK J. CILLUFFO
TASK FORCE CHAIR: SARAH A. MULLEN
TASK FORCE DIRECTOR: LINNEA P. RAINE

About CSIS

The Center for Strategic and International Studies (CSIS), established in 1962, is a private, tax-exempt institution focusing on international public policy issues. Its research is nonpartisan and nonproprietary.

CSIS is dedicated to policy impact. It seeks to inform and shape selected policy decisions in government and the private sector to meet the increasingly complex and difficult global challenges that leaders will confront in the next century. It achieves this mission in three ways: by generating strategic analysis that is anticipatory and interdisciplinary; by convening policymakers and other influential parties to assess key issues; and by building structures for policy action.

CSIS does not take specific public policy positions. Accordingly, all views, positions, and conclusions expressed in this publication should be understood to be solely those of the authors.

Library of Congress Cataloging-in-Publication Data
CIP information available upon request

©1998 by The Center for Strategic and International Studies
All rights reserved.
ISBN: 0-89206-334-3

The CSIS Press
Center for Strategic and International Studies
1800 K Street, N.W., Washington, D.C. 20006
Telephone: (202) 887-0200; Fax: (202) 775-3199
E-mail: books@csis.org
Web site: http://www.csis.org/

Acknowledgments

The Nuclear Black Market Task Force of the CSIS Global Organized Crime Project gratefully acknowledges the contributions of all who participated in this two-day simulation of a nuclear terrorist threat against the United States and those agencies and organizations without whom the work could not have been accomplished. They are the Arms Control and Disarmament Agency; the Central Intelligence Agency; the Departments of Energy, Defense, State, Transportation, and the Treasury; the Environmental Protection Agency; the Federal Bureau of Investigation; the Federal Emergency Management Agency; the Nuclear Regulatory Commission; the U.S. Customs Service; the U.S. Public Health Service; Battelle Pacific Northwest National Laboratory; Lawrence Livermore National Laboratory; Los Alamos National Laboratory; the National Defense University's National War College and Industrial College of the Armed Forces; the U.S. Senate Permanent Subcommittee on Investigations; Hicks and Associates; Ogden Environmental and Energy Services Inc.; and Science Applications International Corporation.

We are indebted to James Schlesinger, former defense secretary, energy secretary, and CIA director; our project chair, William Webster, former CIA director and FBI director; John Holum, ACDA director; Fred Iklé, former defense under secretary; and Gen. Edward (Shy) Meyer, former army chief of staff. Acting the parts of the president of the United States and his most senior advisers, they guided the crucial cabinet-level simulations. We are also grateful to the approximately 70 current and former U.S. government officials and private-sector experts who completed the president's national security team, staffed the interagency working groups, or worked behind the scenes to control the exercise. Their extensive experience with efforts to secure nuclear weapons and materials in the former Soviet Union, combat nuclear smuggling, and prevent nuclear terrorism added immeasurably to the insights gained from this exercise. Most especially we wish to thank the architect of Wild Atom, Daniel Wagner, whose scenario, control of the exercise, and postsimulation drafting were essential to our success.

Finally, we would also like to thank the National Defense University and the Institute for Defense Studies' War Gaming and Simulation Center for cohosting the game. We especially appreciate the students of the National War College class of 1997 whose thoughtful and energetic participation in the exercise dress rehearsal ensured that it was ready for prime time.

The work of this task force could not have been completed without the generous financial support of the Sarah Scaife Foundation, the John M. Olin Foundation, and the Shelby Cullom Davis Foundation.

CSIS also gratefully acknowledges the counsel of the Global Organized Crime Steering Committee, whose advice and scrutiny of this latest task force initiative added an important dimension to the scope and nature of the simulation.

The opinions, conclusions, and recommendations expressed or implied in this report are solely those of the Global Organized Crime Project and do not necessarily represent the views of those organizations just acknowledged.

Contents

Acknowledgments   v

Contents   vii

Project Membership   ix

Remarks by Sam Nunn and James R. Schlesinger   xiii

Foreword   xvi

Key Findings and Recommendations   xix

1. Introduction   1

2. The Setting   3

3. Playing the Game—Day One   7

4. Playing the Game—Day Two   24

5. Assessing The Game—A Few Observations   42

Appendix A.   Wild Atom Participants   53

Appendix B.   Chronology   56

Project Membership

Global Organized Crime Steering Committee Membership

Chair

William H. Webster
Former Director, Central Intelligence Agency
Former Director, Federal Bureau of Investigation

Director

Arnaud de Borchgrave
Senior Adviser, CSIS

Deputy Director

Frank J. Cilluffo
Senior Analyst, CSIS

Members

Duane Andrews
Former Assistant Secretary of Defense
(Director C^3I)

Zoe Baird
Yale Law School

Robert Bonner
Former Administrator
Drug Enforcement Agency

William S. Cohen
Former U.S. Senator
(currently serving as Secretary of
  Defense)

Charles Connolly
Merrill Lynch & Co., Inc.

John Deutch
Former Director
Central Intelligence Agency

Robert Gates
Former Director
Central Intelligence Agency

Carol Hallett
Former Commissioner
U.S. Customs Service

Admiral James R. Hogg
U.S. Navy (Ret.)

Fred C. Iklé
Former Under Secretary of Defense

Stuart Knight
Former Director
U.S. Secret Service

Jon Kyl
U.S. Senator

Walter Laqueur
Cochair, International Research
*   Council*
CSIS

Patrick J. Leahy
U.S. Senator

Bill McCollum
U.S. Representative

General Edward C. Meyer
U.S. Army (Ret.)

Sam Nunn
Former U.S. Senator

Oliver Revell
Former Associate Deputy Director
Federal Bureau of Investigation

William V. Roth, Jr.
U.S. Senator

Donald Rumsfeld
Former U.S. Representative
Former Secretary of Defense

James R. Schlesinger
Former Secretary of Defense
Former Secretary of Energy
Former Director, Central Intelligence
*   Agency*

William Sessions
Former Director
Federal Bureau of Investigation

Admiral William Smith
U.S. Navy (Ret.)

Lieutenant General Edward Soyster
Former Director
Defense Intelligence Agency

J. Chips Stewart
Booz-Allen & Hamilton Inc.

Richard Thornburgh
Former U.S. Attorney General

Curt Weldon
U.S. Representative

R. James Woolsey
Former Director
Central Intelligence Agency

William Zeiner
MITRE Corporation

Global Organized Crime
Nuclear Black Market Task Force Membership

Project Director

Arnaud de Borchgrave
CSIS

Task Force Chair

Sarah A. Mullen
Arms Control and Disarmament Agency

Task Force Director

Linnea P. Raine
Visiting Fellow, CSIS

Task Force Members

George Anzalon
*Lawrence Livermore National
    Laboratory*

Thomas Blankenship
Federal Bureau of Investigation

Edwina Campbell
National Defense University

Burrus M. Carnahan
*Science Applications International
    Corporation (SAIC)*

Paul Dembnicki
Federal Bureau of Investigation

Robert E. Ebel
CSIS

Richard Galbraith
U.S. Customs Service

John D. Immele
Department of Energy

David Kay
Hicks and Associates

Stephen Mladineo
*Battelle Pacific Northwest National
    Laboratory*

Mark Mullen
Los Alamos National Laboratory

Tom Smith
*Lawrence Livermore National
    Laboratory*

John Sopko
Department of Commerce

John B. Stewart
*Ogden Environmental and Energy
    Services, Inc.*

Daniel Wagner
Central Intelligence Agency

George Walker
Federal Bureau of Investigation

Commander Peter I. Wikul (USN)
Joint Chiefs of Staff

Joe Yardumian
Nuclear Regulatory Commission (Ret.)

Observers

Michael Bopp
Senate Permanent Subcommittee on
  Investigations

W. Richard Burcham
Sandia National Laboratories

William E. Clark
U.S. Public Health Service

James A. Genovese
Edgewood Research, Development,
  and Engineering Center

Renee Pruneau
Nonproliferation Center

Remarks

by Sam Nunn and James R. Schlesinger

I salute "National Security Adviser" and "President" Jim Schlesinger for his stalwart efforts to lead his team in devising policies to manage the nuclear terrorism crisis simulated in Wild Atom. In real life, Dr. Schlesinger has led three of the critical organizations involved in this simulation—the Departments of Defense and Energy and the Central Intelligence Agency. Few individuals possess his breadth of U.S. national leadership experience and skills. The problems he and his national security team encountered during Wild Atom point out the horrendous complexities of dealing with the threat of nuclear, chemical, and biological terrorism.

In reviewing the game play and postgame analysis, I was struck by the fact that even these seasoned warriors were uncertain about their options for effective response and frustrated by the slim odds of successful interdiction. I was also struck by the report's observations about the need to integrate the capabilities of our national security establishment with those of our domestic agencies, as well as by the report's suggestion that our National Command Authority may be outdated and should be reviewed. In recent years, I have repeatedly emphasized that our greatest threat—dealing with the increasing availability of weapons and materials of mass destruction to terrorist groups and rogue states—is the threat that we are least prepared for. We must forge closer partnerships between our national security and domestic agencies and devise an integrated, nationwide strategy to improve the capacity of regional, state, and local communities to both prevent and manage the consequences of terrorism involving nuclear, chemical, or biological weapons.

On the plus side, the game demonstrated that our national leadership has made considerable progress in addressing areas of U.S. vulnerability, especially in law enforcement and intelligence and in the development and implementation of scientific and technical countermeasures. The issuance of a Presidential Decision Directive outlining federal responsibilities in preventing and managing terrorist threats, the Nunn-Lugar-Domenici provisions of the National Defense Authorization Act for 1997, and other initiatives are valuable first steps in improving domestic preparedness for events such as depicted in Wild Atom. But much remains to be done.

I applaud CSIS, the Global Organized Crime Project leadership, and the Nuclear Black Market Task Force for taking the initiative in setting up a structured environment, in the form of the Wild Atom simulation, to provide a mechanism for addressing the tough public policy issues related to weapons of mass destruction terrorism. I join the authors of this report in hoping that others will use the experience of Wild Atom to raise awareness about these difficult problems. I also share their hope that Wild Atom will serve as a model for simulation of other types of weapons of mass destruction attacks against the United States.

Sam Nunn
Former U.S. Senator

The end of the Cold War has resulted in a shift of attention from major nuclear exchanges to new, though lesser, threats in a world far less stable. As the magnitude of the threat has diminished, so have the disciplines—including fairly rigorous secrecy—imposed by the superpowers and their allies. Technologies, once highly classified, have been showcased by the Gulf War and have increasingly become more widely available.

Wild Atom deals with one such issue—the possible use of nuclear weapons by subnational groups. This threat has significantly increased with the end of the Cold War. It reflects the growing availability of technical information—and the feared availability of material and manpower consequent on the collapse of the Soviet Union.

Though artificial like all war games, Wild Atom does point to certain vulnerabilities, and, though more suggestive than predictive, the scenario points to important truths. First, the United States, as the world's principal stabilizing power, is a *natural target* for those who are dissatisfied with the status quo. Second, at this juncture, the United States is ill prepared to deal with the problems posed by weapons of mass destruction. Reflect on the massive explosion at Khobar Barracks, clearly an attempt to punish the United States or preferably to help persuade the United States to remove its forces from the Middle East. Consider the impact if a weapon of mass destruction had been available. Alternatively, for a target within the United States, consider how ill prepared we are to deal with such a threat. Recall the panicky reaction to the reactor accident at Three Mile Island—even though that reaction was exaggerated, any such circumstance would certainly attract similar media attention.

In a sense, there is nothing new about such concerns. In the period after World War II, there were fears that the spread of nuclear weapons would be immediate and steady. Among others, President Kennedy speculated on the prospective arrival and impact of 15 nuclear powers. It should be emphasized that, in reality, we did far better than we had feared in slowing the spread of nuclear weapons. But that was in part a reflection of those Cold War disciplines.

Today, we must review those earlier concerns. In the 1960s, a RAND Corporation project (for which I was the project leader) acknowledged the relative porousness of U.S. borders to the introduction of a nuclear weapons. In 1971, at the time of the Cannikin Test, I, as chairman of the Atomic Energy Commission, received a threat to detonate an atomic bomb unless the prospective test was canceled. In those days, given the very limited knowledge of technology and the limited availability of nuclear material, it was easy to dismiss such a threat as a crude bluff. Moreover, in the conditions of that era, it was unnecessary to make such a threat public—and thereby avoid needless media attention or alarm. Today, of course, conditions are quite different.

How should the United States as a nation cope with these changed conditions? We need to be watchful but also to be thoughtful. It is important *not* to overstate the threat. We possess a powerful military posture for deterring most nations and groups that might come into possession of a nuclear weapon. The problem in America today, however, is not panic but rather complacency. We need also to reflect on measures to diminish our vulnerabilities.

Wild Atom makes clear several essential points. First, the danger of nuclear threat has now risen significantly. The availability of special nuclear materials is both substantial and growing—and not just from the former Soviet Union. Restraints have weakened, and technology is more widely available. Second, if a group were to obtain a device, shipping it into the United States might be comparatively easy—indeed, all too easy. We need to devise better measures for border protection. Third, even if some group (or nation) were in a position to cast such hints or to make such threats, if that group were sensible, it would still not *detonate* a weapon. The risks of doing so are very large.

Finally, we need to think through—in advance—effective methods of deterrence. Indeed, how does one deter a terrorist group bent upon a catharsis of violence—one without territory or a population at risk? Clearly one does so, first, by taking steps to ensure that, if anyone were so unwise as to detonate a weapon, the punishment would be both swift and condign.

James R. Schlesinger
Former Secretary of Defense
Former Secretary of Energy
Former Director, Central Intelligence Agency

Foreword

☐ In December 2000, at Russia's Chelyabinsk nuclear-weapons complex, an employee helps the local mafia steal 200 kilograms (kg.) of weapons-grade, highly enriched uranium (HEU) and 20 kg. of weapons-grade plutonium—enough material to fashion two uranium and three plutonium weapons of 10–20 kiloton yield.

☐ The HEU is bought by embittered Chechen separatists, who had been secretly preparing a facility to build nuclear explosives since the suppression of their mid-1990s revolt. The plutonium, purchased by Iran, is stolen on arrival in Tehran by a radical faction of the Iranian-supported Hizballah, which had been seeking fissile material since setting up its covert nuclear facility in Lebanon's Bekaa Valley in 1995.

☐ In early February 2001, less than a month after the new U.S. president is sworn in, U.S. Central Intelligence Agency (CIA) sources report the Hizballah faction intends to mount nuclear attacks against the United States and key allies; but intelligence analysts question Hizballah's capabilities and whether the group was acting alone or at Iran's behest.

☐ Chelyabinsk officials, discovering the theft while preparing for U.S.-assisted security upgrades, confine in a hotel a group of visiting Los Alamos National Laboratory technicians (who alert Washington to the crisis atmosphere); and the National Security Agency (NSA) detects unprecedented activity on Russia's highest-level secure communications.

☐ Two days later a nuclear detonation near Moscow is detected by a U.S. satellite. Russian authorities are overwhelmed, coping with the consequences of what appears to be an accident. President Lebed of Russia places a call to the U.S. president and states that the incident did not involve a government movement of weapons or other nuclear-related material. The Chechens claim credit for the event, and both the Chechens and Hizballah threaten nuclear attacks (against Russian, European, and U.S. interests) unless Russian troops withdraw immediately from Chechnya and U.S. forces withdraw from the Arab world.

☐ Meanwhile, after New York City harbor police arrest an Iranian student whose laptop computer contains detailed reports and sketches of Atlantic port cities (already e-mailed via the Internet to a private Beirut company), the Federal Bureau of Investigation (FBI) activates the Federal Radiological Emergency Response Plan and the Department of Energy (DOE) deploys

Nuclear Emergency Search Team (NEST) technical experts to the FBI's field office.

❏ Unknown to all, a Greek freighter that is carrying a device Hizballah plans to detonate while the ship is in Baltimore harbor is due to arrive in port on February 14.

Rapidly unfolding events confronted the president in the simulated nuclear terrorist incident, code-named Wild Atom and staged by the Center for Strategic and International Studies (CSIS) in conjunction with National Defense University.

Wild Atom demonstrated that the United States is singularly ill-prepared to deal with an act of terrorism that involves weapons of mass destruction (WMD). Much of what needs to be done will take several years to accomplish—well beyond 2001. Major obstacles must be overcome if we are to equip people with rudimentary protection against an act of nuclear terrorism on U.S. soil.

The exercise also revealed a wide gap in roles and missions between domestic agencies and the national security apparatus. The national security mandate needs to be broadened to include domestic activities and players.

This report makes recommendations about what can realistically be achieved to minimize risks in both the near term and the long term.

The United States is still not equipped to manage the threat of WMD against civilian populations, let alone the use of these weapons or the consequences of large-scale devastation. The first tentative efforts to train first responders such as state and local police, fire fighters, emergency service providers, and medical personnel—in 120 cities eventually—have proved woefully inadequate.

Congressional hearings in October 1997 chaired by Representative Curt Weldon (R-Pa.) revealed that not a single cabinet-level official in the current administration has ever participated in a WMD exercise. Sadly lacking, too, is congressional support for a U.S. response to the threat of WMD terrorism that is similar in spirit to America's vigorous Cold War strategy. Senator Richard Lugar (R-Ind.) and Rep. Weldon have stated publicly that Americans have every reason to anticipate acts of nuclear, chemical, and biological terrorism against domestic targets before another decade is out.

Greater efforts at domestic preparedness also have a critical international dimension—better intelligence capabilities and intelligence sharing with like-minded nations.

In the interval between the simulated nuclear terrorist incident Wild Atom and the publication of this report, a journalistic couple published the book *One Point Safe* in which an entire chapter purportedly described what transpired during the CSIS game. The book, in turn, was the inspiration for the movie *The Peacemaker*. The film triggered a debate about the merits of educating the American public about the threat. The idea that popular culture has no role in nuclear security issues is wrong. As Rep. Weldon put it, the movie was a disturbing case of art imitating life and many premises of the motion picture are based on grim realities.

The report *Russian Organized Crime* released in October 1997 by the CSIS Global Organized Crime (GOC) Project documented that corruption and organized crime in the Russian military have reached such a scale that they have endangered the security of nuclear weapons and materials. And one need only read the daily headlines of Iraq's latest interference with UN inspections or details of Iran's continuing pursuit of a nuclear-weapon capability to appreciate that customers exist for smuggled nuclear weapons and materials that could have horrific consequences in the hands of rogue states or terrorists.

Whether the happy ending portrayed in *The Peacemaker* is the likely outcome of an actual nuclear terrorist event "is highly problematical," according to Rep. Weldon. The Wild Atom exercise had no actual ending, happy or otherwise; it came to a halt with the conclusion of the first National Security Council (NSC) meeting of America's new president in 2001. But we hope that the exercise, and the report it generated, can serve as a model for similar activities involving the nation's most senior officials.

Wild Atom: Nuclear Terrorism is the fourth in a series of CSIS Global Organized Crime Project publications. By bringing together multidisciplinary expertise (such as that residing in the Nuclear Black Market Task Force), the GOC Project has provided a vehicle to develop short- and long-term strategies and mechanisms to better understand and deal with national security threats posed by transnational crime. The Nuclear Black Market Task Force is only one of the seven separate task forces that make up the project. Like the Nuclear Black Market Task Force, each focuses on specific sets of issues and threats with a view to developing more precise policy, technological, and organizational recommendations.

Judge William H. Webster
Project Chair

Arnaud de Borchgrave
Project Director

Key Findings and Recommendations

Few counterterrorist experts today doubt the appeal of nuclear terrorism to ruthless terrorists like those who carried out the vicious attacks in recent years in New York, Oklahoma City, and Tokyo. Nuclear explosives are no longer beyond the reach of terrorists: the technology is a half-century old, and weapons-usable fissile material is available on the black market. Should terrorists acquire a nuclear device, they could use it to intimidate, extort, cause panic, contaminate, or destroy.

Lessons learned from the unclassified Wild Atom simulation address the key issues of incident management (especially after a nuclear incident), intelligence, policy, and international cooperation.

Incident Management

If determined terrorists were to gain possession of a "loose nuke" today, the knowledgeable participants in Wild Atom concluded, successful interdiction would be extremely difficult. If their judgment is correct, the United States must immediately take steps to deal with a nuclear detonation or radiological dispersal on U.S. soil. The United States should

❑ Improve consequence-management preparations at home and in cooperation with close allies. U.S. civil defense measures are improving, but what of all the other preparations necessary to cope with the effects of and clean up after a nuclear detonation or contamination? In particular, the U.S. national security and national disaster medical communities need to work together more closely.

❑ Improve U.S. and allied capabilities to neutralize a nuclear device. The United States has some impressive capabilities, but could it work effectively with other states to locate and disarm a nuclear bomb? If not, are new forums, agreements, information sharing, and joint exercises needed? Have U.S. officials resolved the political and legal implications of a foreign request for help in disabling and disposing of a nuclear weapon?

Intelligence

Whatever chance there is to interdict a nuclear weapon coming to America, participants concluded, depends heavily on the availability of actionable intelligence. In addition to continued high-priority efforts to penetrate terrorist groups and proliferant states, Washington should also promote

❏ Training and exercises to develop substantive expertise and effective working relationships among counterproliferation and counterterrorism analysts and operators on the one hand and appropriate officials in the policy, defense, law enforcement, and technical communities on the other.

❏ Innovative international arrangements to share information promptly and to craft trafficker profiles jointly, identify patterns in the activities of nuclear criminals, determine the origin and route of seized nuclear materials, and assess the bomb-making capabilities of key rogue states and terrorist groups should they acquire sufficient fissile material. The United States needs to move beyond traditional bilateral intelligence liaison toward a multilateral pooling of information, analysis, and resources.

❏ Increased funding and a sense of urgency to develop new sensors to detect both uranium and plutonium (and possibly other radioisotopes). The improved detectors need to be effective at a distance and despite shielding around the fissile material. They should be mobile, concealable, easy to operate, and inexpensive.

Policy

While working to improve intelligence and detection capabilities, the United States should also improve its policy process, Wild Atom participants urged.

❏ Top officials must curb bureaucratic rivalry and parochialism to inculcate a culture of cooperation and a national strategy. Intelligence and law enforcement agencies are cooperating, but they need to do more. Other agencies that handle domestic U.S. emergencies or that could gain needed information from foreign counterparts must join the traditional national security agencies in this effort.

❏ Before the heat of a crisis, U.S. policymakers need to exercise response procedures, not only to ensure interagency cooperation but also to think through key issues, for example, guidelines for dealing with nuclear terrorists. Negotiations might avert a nuclear disaster, but they might also encourage future attempts at nuclear extortion.

❏ Other key crisis response questions must be similarly addressed. Officials should try to establish the conditions that would justify search techniques that disrupt commerce and traffic. More important, what would justify evacuation of a U.S. city?

Dissatisfied with current policy, one senior participant suggested that the United States should adopt a tough, "you will die if you try" stance toward nuclear terrorists. He suggested three steps:

❑ *Defense:* In the event of a terrorist device coming to America, close U.S. borders temporarily and push the nuclear detection effort offshore.

❑ *Deterrence:* Declare publicly that any entity possessing nonsafeguarded nuclear material must give it up or be considered fair game for U.S. preemptive action. The United States should also increase rewards for surrendering fissile material or providing information on nuclear traffickers.

❑ *Response:* Eliminate any entity that causes a nuclear explosion. In addition, brand nuclear criminals as international outlaws and invoke a firm policy of relentless pursuit of nuclear terrorists, much as law enforcement agencies today chase down "cop killers."

International Cooperation

Finally, Wild Atom participants urged, the United States should

❑ Promote multilateral regimes that effectively combat proliferation of nuclear weapons and weapons-usable materials. How can we strengthen existing international norms? Do we need a nuclear smuggling convention?

❑ Ensure secure control by responsible government authority over all remaining warheads and fissile materials. Should we move faster to help Russia secure the former Soviet nuclear stockpile? Are we institutionalizing U.S. assistance so that Russia can sustain these programs after our aid ends?

❑ Strengthen international cooperation to try to interlock national defenses into a seamless, layered barrier against the smuggling and use of nuclear weapons and materials. Publicize these interlocking defenses to deter would-be nuclear traffickers. Several U.S. agencies are independently training foreign police, intelligence, and policy officials; but do we need institutionalized multinational arrangements? Because each state will try independently to keep nuclear materials out of its territory, achieving effective cooperation that involves risk taking will require time and perseverance.

❑ Implement a public diplomacy campaign to brand nuclear terrorists as international outlaws. This campaign should include publicizing the severe penalties handed out to anyone convicted of a nuclear crime. It may also include offering bounties and amnesty for help in neutralizing nuclear criminals.

Raising awareness of the nuclear terrorist threat among senior U.S. policymakers may be the first essential step toward strengthening our security. Toward that goal, the CSIS Nuclear Black Market Task Force hopes that Wild Atom will provide a timely alert rather than a frightening glimpse of the future. Although the lack of any confirmed illicit transfers of weapons-usable nuclear materials in the past few years appears to have produced overconfidence in some government circles, experts do not know whether the nuclear black market has been eliminated or only better concealed.

CHAPTER 1

Introduction

Wild Atom: Nuclear Terrorism covers the unclassified Wild Atom exercise conducted in November 1996 by the CSIS Global Organized Crime (GOC) Project's Nuclear Black Market Task Force and the National Defense University's War Gaming and Simulation Center. The report reviews and assesses the exercise and proposes initiatives to strengthen U.S. policy and capabilities against nuclear terrorism. The game represented an important milestone in that it allowed the task force to revisit in an exercise setting its earlier task force report, *The Nuclear Black Market*. Building on that report, Wild Atom has incorporated elements of the topics of nuclear materials smuggling, organized crime, and terrorism.

What *The Nuclear Black Market* analyzed, the Wild Atom exercise simulated, thus providing government officials and substantive experts an opportunity to face the potential challenges and consequences of nuclear smuggling and terrorism and to evaluate U.S. defenses. Designed to exercise rigorously various dimensions of this threat, the Wild Atom exercise had the following objectives:

❑ Assess U.S. capabilities to prevent or respond to nuclear-weapon incidents abroad and at home.

❑ Examine the interplay of intelligence, law enforcement, and scientific-technical crisis support for U.S. national decision-makers.

❑ Explore the state of coordination among the U.S. intelligence, law enforcement, and scientific communities.

❑ Highlight technological constraints in detecting weapons-usable nuclear materials.

❑ Evaluate the adequacy of existing international laws and agreements to prevent nuclear smuggling.

❑ Underscore the potential impact of Russian organized criminals should they become heavily involved in stealing and smuggling nuclear material.

The scenario incorporated plausible perpetrators.

❑ *A thief.* A trusted insider helped organized criminal elements gain access to the facility.

❑ *Traffickers.* A Russian mafia gang located buyers both inside and outside Russia and smuggled the goods to them. (Russian interior ministry officials acknowledge that Russia is rife with criminal gangs, ranging from localized thugs to extensive groups with international connections.)

☐ *Terrorists*. The two buyers were a substate separatist movement (the Chechens) and a state-sponsored international terrorist group (Hizballah). Factions of both groups possess the motives and means to carry out a terrorist attack on today's most deadly scale.

Although the terrorists varied in their willingness actually to use a nuclear device, each group was ready at least to threaten a nuclear attack. They also shared a conviction that an actual device was necessary to make their threats credible and that two or more devices would be needed because a demonstration might be required.

For added realism and context, the exercise postulated a theft of fissile materials from an existing ex-Soviet facility in present-day Russia. The materials that the Wild Atom exercise simulated being stolen are actually stored at that facility today in quantities and at enrichment levels sufficient to build nuclear explosives. In Wild Atom, the terrorists acquired both uranium and plutonium, which pose distinct challenges for weapons construction and detection. Thus the exercise involved two weapon designs: a simple (uranium) gun-type device and a more sophisticated (plutonium) implosion weapon.

The terrorists could choose among several targets: the U.S. homeland, U.S. troops and other interests overseas, and allies and partners of the United States. Delivery was to be by improvised means: a truck or ship that terrorists could easily use to accommodate a crude and therefore bulky device. In planning to attack a U.S. port city (Baltimore), the terrorists intended to detonate the device aboard ship. That eliminated any need to unload components past U.S. customs controls. It also avoided the radiation detectors that the U.S. Customs Service announced in June 1996 it was deploying at some U.S. airports and ports of entry.

Finally, the written inputs and structure of the exercise faithfully replicated all-source intelligence practices, U.S. bureaucratic rivalry, and a crisis atmosphere. The meetings and documents followed actual government practice, including intelligence summaries, working-group meetings, briefings of U.S. principal officers, and sessions of the cabinet-level Principals Committee and the president-chaired National Security Council (NSC).

Chapter 2 provides the detailed scenario on which the game is based; Chapters 3 and 4 review game play during Days One and Two; and Chapter 5 assesses the results of the two days of play.

The Setting

It is February 2001.

> *In Washington, the new president is a veteran of many years in the Congress. He is widely regarded as knowledgeable of international affairs, is decisive and securely in control of his foreign policy team, and takes pains to cultivate public and congressional support for his handling of foreign policy.*

> *In Moscow, President Aleksandr Lebed has been in office for two years following President Boris Yeltsin's death. Economic distress, official corruption, and organized crime remain pervasive. Separatist sentiments in ethnic minority areas of Russia remain strong. In particular, Moscow's forces continue their brutal occupation of Groznyy, while the underground Chechen leadership continues to prepare for another uprising.*

> *In the Arabian Peninsula, U.S., British, and French forces still dare not leave for fear of renewed Iraqi aggression. Organized terrorist groups and other thoroughly radicalized Muslims chafe at the continuing U.S.-led presence.*

Otherwise, the world remains largely as we know it.

The Theft

The mafia boss in Chelyabinsk considered the proposition: For a large cash payment and an alias passport, the nuclear engineer offered to turn over a fortune in weapons-usable fissile material. Until the implementation of U.S. security measures—in two months time—at the new fissile material storage site, the engineer could bypass existing security. There for the taking were 200 kg. of weapons-grade, highly enriched uranium (HEU) and 20 kg. of plutonium.

The boss knew that Chechnya would pay handsomely. The Groznyy underground for years had been preparing a secret facility safely outside Chechnya (actually located in the Russian city of Rostov) where they could build a nuclear device, and the separatists had been hounding all the Chechen-run gangs to steal a warhead or some of this precious material. Moreover, Chechnya could not afford all of this huge haul, so the rest could be sold to another buyer. Indeed, the demand side of the nuclear black market had become increasingly active and insistent since the late 1990s.

There was a downside to the operation, however. Despite the Russians' notoriously inept accounting, such a large theft would be detected when the material was inventoried and the security upgrades installed. By that time, everyone involved would have to be permanently outside of Russia. The local authorities would be furious that the mafia had broken the agreement that condoned black market dealings in the closed city in exchange for leaving the nuclear facilities alone.

His mind made up, the boss gave orders to contact the potential buyers immediately. The Chechen gang had well-established international connections after years of smuggling contraband, and communications were reliable and quick.

Knowing that the HEU permitted simpler bomb designs and was less detectable than plutonium, the gang offered the uranium to Groznyy. Agreement was immediate. In turn, Tehran was offered the plutonium. Iran agreed to pay in full upon delivery. Neither buyer knew of the other sale.

The material was stolen, divided, and moved in several vehicles to avoid an unsafe concentration.

The uranium arrived in Rostov the day after the theft, while the plutonium took two weeks to move along an established smuggling route through the Caucasus to Iran. On delivery, Iran arranged final payment to a bank in Zurich via electronic transfer from an expatriate Lebanese account in Detroit, and Chechnya arranged several payments from overseas patrons of the Chechens. The Chelyabinsk insider vanished, and the gang members slipped into Azerbaijan, whence they dispersed.

The theft was detected nearly two months after the fact on the eve of the U.S. technicians' arrival. Local Russian officials halted the security upgrade and informed the Ministry of Atomic Energy (MINATOM) and the Ministry of Defense (MOD). Moscow immediately mobilized Ministry of Interior (MVD) nuclear security units, general staff elements, MOD special forces (*Spetznaz*), and the FSB's special Alpha unit of the Federal Security Service (FSB); but it was too late to find the material or the thieves.

The Buyers

Since their first failed uprising, the Chechen separatists had wanted a nuclear capability to deter President Lebed from sending troops to crush a second rebellion. They believed that would force Russia to accept an independent Chechnya and not interfere with the state's external relations and trade. Groznyy had set up the Rostov covert facility and equipped it with experienced weapons experts, tools, and all the nonfissile components. Arrival of the HEU triggered a crash program to manufacture two bombs, which were ready in six and eight weeks, respectively. The experts insisted the crude gun-type devices were reliable even though untested.

The Chechens dispatched their bombs to Kursk and Moscow. When all was ready, they planned to threaten publicly to destroy Russian and Western cities if Russian forces did not promptly leave Chechnya. No one actually intended to detonate a device, but a few hard-liners believed it might be necessary. If absolutely forced, the Chechen leaders conceded that they might explode the device outside Kursk and threaten that a city—even Moscow—would be next. If Lebed still refused, the hard-liners argued, Chechnya could bring irresistible domestic Russian and Western pressure to bear by exploding the Moscow device and announcing that more bombs were ready to detonate in various European cities.

The Iranians were equally determined to acquire a nuclear capability, but they were even more divided over how to use it. A moderate majority insisted that a credible nuclear threat would suffice to evict U.S. and other Western forces from the Middle East. Anticipating that Western sanctions would continue to hobble Iraq, Iran believed that it could dominate the Arabian Peninsula once the foreign troops had departed. Relying on this bomb-in-the-basement logic, Tehran would hide the fissile material and defer any actual decision to assemble a device. A radical minority, however, saw this plan as inadequate; the radicals wanted to assemble as many devices as possible. They also wanted to detonate one to punish the United States, compel a humiliating U.S. withdrawal, and leave no doubt of Iran's status as the regional hegemon. If Iranian president Khatami lacked the stomach for such audacious action, the radicals knew that Hizballah did not.

The Iran-backed Hizballah terrorist group had, in 1995, set up a covert facility in Lebanon's Bekaa Valley where it was making chemical weapons. At the secret site, the group also had assembled the wherewithal to build a nuclear weapon; meanwhile, it waited for Iranian sponsors or smuggling contacts in Russia to provide the essential fissile material. Told by sympathizers within the Iranian government that Tehran was about to receive such material for its own use, Hizballah immediately began plotting to divert it. Other acquisitions were accelerated, including the recruitment of two disgruntled, out-of-work, former Soviet nuclear-weapon experts. Desperate to prove their bona fides, the experts had already been advising the Hizballah procurement efforts and even providing workable weapon designs via the Internet.

To aid several of their necessarily covert purchases, Hizballah—with official Iranian help—had established a front company in Tehran. Unknown to the Iranian government, however, the firm used its connections with Russian smugglers to contact a small company just outside Pelindaba, South Africa. The owners of the plant were ex-employees of Adventa Central Laboratories, once part of the apartheid state's now-dismantled nuclear-weapons program. Advised by their Russian experts that three implosion devices could be made from the 20 kg. of plutonium, they placed an order with the South Africans for the exotic triggers and chemical explosives needed to compress the plutonium into a supercritical mass during detonation.

Finally, Hizballah planned how to use the weapons. To transport the crude and necessarily large devices securely, they planned to use ships or large trucks. Each bomb would be dispatched as soon as it was produced. Targets included the U.S. forces in Saudi Arabia plus the United States, England, and France. Delivery would be by sea out of Beirut. The first device would go to a U.S. harbor, the second to a port in England or France, and the third to the port city of Dhahran, Saudi Arabia.

The United States was the first and primary target. A Hizballah agent in the United States—an Iranian student—was instructed to inspect major ports along the eastern seaboard. He was to look for a large harbor close enough to a major city so that an explosion aboard ship would inflict massive blast casualties. To spread radiation for maximum casualties, the target should have prevailing onshore winds. Just two days later, the agent began e-mailing reports by laptop computer via the

Internet to a private company in Beirut. Meanwhile, the first device, completed in late January, was moved to Beirut and loaded into the hold of a small Greek freighter that regularly moved cargo back and forth across the Atlantic Ocean. The nondescript vessel would draw no special attention, and detonating the device aboard ship in the harbor would avoid having to unload it and move it through customs inspection. Work continued on the other two plutonium weapons.

Playing the Game—Day One

NATIONAL SECURITY COUNCIL
Washington, D.C. 20506
February 10, 2001

MEMORANDUM FOR

MR. EDWARD BLAKE
Assistant to the Vice President
for National Security Affairs

MR. THOMAS JORDAN
Executive Secretary
Department of Energy

MR. JAMES SMYTHE
Executive Secretary
Department of State

MR. FRANK BROWN
Executive Secretary
Central Intelligence Agency

MS. JANICE DANFORTH
Executive Secretary
Department of Treasury

CAPTAIN B. L. KESSING
Secretary
Joint Chiefs of Staff

COL MICHAEL RAWLINGS
Executive Secretary
Department of Defense

MRS. JACLYN WEST
Office of the Representative
of the United States
to the United Nations

Subject: Principals Committee Meeting on the Missing Russian Nuclear Materials
and Iran/Hizballah.

There will be an expanded Principals Committee Meeting on the Missing Russian
Nuclear Materials and Iran/Hizballah in the White House Situation Room, February
10, 2001, from 6:00 P.M.–8:00 P.M. Attached at Tab A is the agenda for the subject
meeting. Attached at Tab B is a CIA assessment that will provide the basis of the
discussion.

William B. Goode
Executive Secretary

Attachments
   Tab A   Agenda
   Tab B   CIA Assessment

Tab A

PRINCIPALS COMMITTEE MEETING ON
THE MISSING NUCLEAR MATERIALS IN RUSSIA
DATE: February 10, 2001
LOCATION: White House Situation Room
TIME: 6:00 P.M.–8:00 P.M.

AGENDA

I.   Opening Remarks   NSC

II.   Intelligence Update   DCI

III.   Results of IWGs   DCI, Secretary of Energy, Attorney General

IV.   Policy Discussion   All Participants

    A. Presidential Action—does the situation warrant the President demand-
        ing an explanation from President Lebed?

    B. Preemptive Action

        —Increase security of U.S. forces in Saudi Arabia?

        —Offer assistance to Russia or other nearby states to recover the miss-
          ing fissile material?

        —Take steps to locate and neutralize any Iranian or Hizballah nuclear
          capability?

        —Covert Action?   —Military?   —Unilateral?   —Multilateral?

        —Implement additional security measures in the U.S.?

        —Assist the threatened European Allies?

    C. Congressional Notification?

    D. Press Guidance.

Tab B

CENTRAL INTELLIGENCE AGENCY
Washington, D.C. 20013

DCI Counterterrorism Center
10 February 2001

Hizballah May Be Preparing Nuclear Attack on United States

Executive Summary

The Iranian-backed Hizballah terrorist group reportedly has assembled several workable nuclear weapons and intends to attack urban targets in the United States, UK, or France as well as U.S. armed forces in Saudi Arabia, according to an untested human intelligence source with proven access to senior Iranian Government officials in Tehran. Hizballah, which may be under Iranian Government control or may be conducting a rogue operation not sanctioned by Tehran, obtained the material several weeks ago from Russia.

In Russia, secure communications serving the senior political leadership, General Staff, and nuclear units remain unusually active. U.S. nuclear experts from Los Alamos National Laboratory, just arrived in Chelyabinsk for a scheduled security upgrade, suspect that a search for stolen fissile material is under way.

In Washington, agencies vary in their assessments of: (1) the probability that Hizballah possesses fissile material and a means to disperse contaminants (i.e., a radiation dispersal device, or RDD); (2) Hizballah capabilities to manufacture a nuclear explosive (i.e., an improvised nuclear device, or IND) assuming that the group possesses weapons-grade fissile material; and (3) whether Hizballah is acting independently or under Tehran's control. Interagency working groups from the intelligence, law enforcement, and nuclear technical communities are meeting today to consider these questions and to formulate U.S. options. In view of the short time remaining before the urgently scheduled Principals Committee meeting, recommendations from the IWGs will be presented at the table.

Latest Intelligence

The new and untested source of this morning's CIA report could not identify how many devices there are, where they are located, when or how they might be moved to their targets, or the relationship between Hizballah and the Iranian Government. Senior Iranian officials in Tehran were confidentially informed through government channels of Hizballah's claims. The officials appeared to be convinced and

-1-

genuinely concerned that Hizballah has the material and assembled bombs, that the devices are crude but reliable, and that Hizballah is determined to attack cities or ports in the United States, the United Kingdom, and France (all of which have stationed armed forces in Saudi Arabia since the end of Desert Storm) as well as U.S. forces in Saudi Arabia.

Update in Russia

Russian secure communications serving the senior political leadership in Moscow, the General Staff, the nuclear oversight agencies of MINATOM and GAN, MVD nuclear security forces, selected Spetznaz units, and the FSB's nuclear-designated "Alpha" unit have continued to transmit three times the normal message traffic since approximately 2:00 A.M. yesterday, 9 February, Washington time. Outside Moscow, most activity is concentrated in the southern Ural Military District near the Kazakhstan border, and also in the Transcaucasus around Chechnya. Russian border guards along the Kazakh border are on alert, as are border units in the Caucasus.

- NSA convened a NOIWON telephonic alert at 05:20 A.M., 9 February. Four hours later, the President placed a hot-line call to Moscow. President Lebed was reassuring but uninformative.

Possible Nuclear Theft. At 4:00 A.M. this morning, 10 February, TDYers from Los Alamos National Laboratory at the Russian Chelyabinsk nuclear-weapons complex reported by satellite telephone that they were being confined to their hotel without explanation. The TDYers arrived yesterday to help their Russian counterparts begin inventorying fissile materials and upgrading security measures at the facility. They described what seems to be a complete lock-down of the facility and heavily reinforced perimeter security. The U.S. technicians themselves observed extensive military traffic through the town, and hotel employees say that a search is under way for "local mafia." Although their instruments gave no indication of a nuclear release, the U.S. technicians speculated that an actual or simulated nuclear theft might be involved. They reported that the Chelyabinsk site reportedly contained no nuclear weapons, but weapons-usable uranium and plutonium had been consolidated there for secure storage.

Developments in Iran

The reports out of Russia may shed light on otherwise unexplained developments in Iran. For several weeks, a political crisis had been building in Tehran.

- President Khatami's ruling Socialist Party nearly six weeks ago stepped up media attacks on Hizballah. The U.S. Interests Section in Tehran reports that several Iranian Government officials known to sympathize with the terrorist group have been arrested for "theft of state property." Several top Hizballah officials are rumored to have escaped to Lebanon.

- CIA sources report that a smuggled shipment of military goods from Russia was stolen on arrival more than a month ago in Tehran. The U.S. Embassy advises that Khatami's government is going to extraordinary lengths to locate and recover the shipment.

- NSA analysts searching recent intercepts on Iranian commercial activity for clues to the missing shipment discovered an urgent order to a firm in South Africa for what analysts believe to be nuclear firing triggers and related weapon components. An Iranian firm believed to be connected to the Iranian nuclear-weapons program and other covert Iranian programs—including aid to Hizballah—placed the order and may have already received the goods. The South African firm is a spin-off of Adventa Central Laboratories, outside Pelindaba, which was part of the apartheid regime's now-dismantled nuclear-weapons program. CIA analysts believe the only logical use the Iranians might have for the components is in a uranium or plutonium implosion device.

Is Tehran In Charge? An intercepted commercial financial transaction reveals a highly unusual overseas payment from an Iranian-affiliated account in the United States in late December, 2000. A bank in Detroit, Michigan, wired $20 million from accounts controlled by the local Lebanese community (with known ties to Tehran) to an account in Zurich. Cross-checking reveals that the Swiss account formerly belonged to a known Soviet KGB front company. Currently, the company is suspected of financing Russian organized criminal activity. That Swiss account the same week received five other large transfers, totaling another $23.5 million, from commercial entities in Europe and Asia. CIA has no information on the additional payers.

- The U.S. Interests Section in Tehran believes this evidence implicates the Khatami government in a purchase of nuclear material stolen in Russia. If so, however, it remains unclear whether the material was purchased by Tehran on behalf of Hizballah, or was purchased for Iran's own purposes but stolen by Hizballah and/or its sympathizers within the Iranian Government. It is equally unclear if the Iranian firm's purchase of probable nuclear triggers was on behalf of Teheran or Hizballah. The unprecedented recovery effort under way these past few weeks in Iran appears genuine (rather than an elaborate deception effort), which suggests that Tehran is indeed trying to recover stolen property.

U.S. Interagency Working Groups Convened

We have provided this assessment to all appropriate agencies and departments in Washington. Our informal contacts suggest that there may be significant differences in agency perceptions of Hizballah and Iranian capabilities and intentions: Accordingly, representatives of the intelligence, law enforcement, and nuclear technical communities are meeting to review the reporting and to formulate U.S. policy options. Recommendations of the IWGs should be available at the Principals Committee meeting this afternoon.

Day One—the ambiguous warning phase—included two rounds of simulated interagency deliberations, the interagency working group (IWG) meetings, and the Principals Committee meeting. The meetings were in response to the announcement by the NSC's executive secretary of the Principals Committee meeting and the Central Intelligence Agency (CIA) report of the possible impending nuclear attack on the United States.

In Round One, participants who played the roles of working-level officials met separately in the Intelligence, Law Enforcement, and Science/Technology IWGs.

Intelligence	Law Enforcement	Science/Technology
Central Intelligence Agency (CIA)	Department of Justice (DOJ)	Department of Energy (DOE)
National Security Agency (NSA)	Federal Bureau of Investigation (FBI)	National Laboratories
Defense Intelligence Agency (DIA)	Department of Transportation (DOT)	Joint Chiefs of Staff (JCS)
Department of Energy (DOE)	U.S. Coast Guard	Defense Special Weapons Agency (DSWA)
Department of State Bureau of Intelligence and Research (INR)	U.S. Customs Service	Nuclear Regulatory Commission (NRC)
		Environmental Protection Agency (EPA)
		Federal Emergency Management Agency (FEMA)

Each participant received a copy of the CIA analysis of the known events in Russia and Iran. The IWGs assessed the situation, took initial actions (such as requesting more information from intelligence and diplomatic sources), and briefed their principal officers. The Intelligence IWG briefed the director of central intelligence (DCI); the Law Enforcement IWG briefed the attorney general and the FBI director; the Science/Technology IWG briefed the secretary of energy and the chairman of the Joint Chiefs of Staff.

Observers noted adequate cooperation in each working group, but bureaucratic rivalries and parochial viewpoints did generate some tension. The groups struggled with a variety of issues:

❑ Too often the agencies that must cooperate and work closely together to prevent an improvised nuclear device from reaching U.S. shores or to clean up afterward appeared unfamiliar with each other's roles and capabilities.

❑ Although some participants in the Law Enforcement IWG denied it, observers noted competition among the FBI, Coast Guard, and Customs Service to take the lead. Valuable time was spent sorting out jurisdictions and priorities. In due course, the FBI took the lead on the criminal investigation and counterterrorism, while the Coast Guard assumed responsibility for interdiction at sea.

❑ The Intelligence IWG was frustrated that the scenario ruled out many standard preemptive measures and that available shorter-term measures were both costly and had an unacceptably low probability of success.

❑ The Science/Technology IWG optimistically assumed detection capabilities that might be available by the year 2001; however, it is generally doubted that the present trajectory of low funding can fuel the discovery and deployment of these technologies.

In Round Two, the president's national security adviser convened the cabinet-level Principals Committee to consider the ambiguous indicators of the developing crisis.

He opened the session with a scene-setter that underscored the potential gravity of the situation:

> **National Security Adviser:** Thank you all for coming. We face a situation that is quite somber and one that perhaps reflects a risk embodied in the nuclear age.
>
> As you all know, for many years the United States has with reasonable consistency followed a policy of attempting to prevent proliferation.... The great irony that we face is that our victory in the Cold War has opened the possibility of a nuclear attack on the United States and on our allies simply because the control and security in the follow-on states to the Soviet Union is not what it might be. It is also true that as the leading power in the world no one is likely to challenge us militarily so that terrorism has become the tool of choice as a way of getting at the United States. And these two aspects have flowed together, possibly to create risks to this country.
>
> I want to see what we are doing with regard to liaison relationships. We need to be very careful in circumstances of this sort (in which each nation will have a temptation to take care of its own needs by shuffling the possibility of nuclear attack onto others) to see to it that we can pull together as well as possible.
>
> The president has been informed about this meeting, and, of course, we will have a meeting of the National Security Council. We all must recognize in this situation that the president was inaugurated just over one month ago. The consequences of this crisis or potential crisis will color his whole administration. The public must see that this president has done all that is possible to prevent an assault on the United States. That means that the President will probably feel the necessity of communicating these problems early on to the American people and that we must have an appropriate information policy with regard to these risks.
>
> Let us turn now to the results of the various working groups. I don't know to what extent you have been working together and having some kind of synergy. But we must be assured that we do not have agencies working independently of other agencies. And we must be assured that we do not have a great deal of suboptimization. We are all working together to assure the best *national* policy not the best *agency* policy. And that is the purpose of this group.

After the national security adviser's opening remarks, the DCI provided his intelligence update and the lead principals briefed on the results of their respective IWG meetings.

The DCI reviewed the available intelligence in timeline fashion. He offered three scenarios to frame the potential threat: most likely was the existence of a crude and unwieldy nuclear device, the worst case was a more sophisticated and compact weapon, and the best case was no device yet assembled:

> **DCI:** If that's the first nuclear material they got, what could they have by now? Our people say, if there is something that explodes today, it's a pretty unwieldy device or it has to be transferred in pieces and reassembled. My worst case is that we are only getting in late in the game and some things have happened well before this. Which means that you have got something that is transportable in a somewhat more compact form and a more sophisticated piece. The best case is that we did have some equipment stolen, that the nuclear material theft is what caused the Russian bells to go off, which means we have some time and we have some more of a geographic concentration to focus on. We are going to be collecting on all three.
>
> **National Security Adviser:** What are we picking up from our allies through our liaison relationships?
>
> **DCI:** We have so far not heard from our allies, but we are going to go to them en masse. It would help us to find out exactly the content of the president's discussion with President Lebed because we are trying to find out what do the Russians know, what's their perceptions, so that we can go back to them.
>
> I should be honest with you, sir. We believe that we've got to go very tight with the Russian service. Time-out for any bad feelings. Everybody is at risk here. We've even had suggestions—subject to policy decisions, of course—that we even go so far as [contacting] the Iranians because it is not clear what stake they have here.
>
> We are operating on the principle until proven otherwise that nuclear material has been stolen. It is in the hands of a threatening group. Our questions are at what stage of the process are they and how do we find out where and what?

The secretary of energy, beginning with the simplest and most likely, further bounded the potential threat in terms of three possible weapons: a radiological dispersal device, an improvised nuclear device, and an implosion device. The director of Los Alamos described the size and weight of each possible device to draw implications for available means of delivery:

> **Secretary of Energy:** There are really three things…one might do with this material to fabricate a device: a radiological dispersal device, improvised nuclear device, or an implosion device.
>
> With respect to an RDD [radiological dispersal device], damage to human health would be a function of the density of the area. In lower Manhattan, that would be catastrophic.
>
> Secondly, with respect to an improvised nuclear device,…Iran and Hizballah have the technical capability to fabricate and detonate a [gun-

type HEU] device of 2–10-kiloton yield. All can do it alone. With respect to an implosion device, you get a yield of 20 kilotons, and the Iranians can certainly do it by themselves. With help, Hizballah can do it. We estimate that it is unlikely that Hizballah can fabricate an implosion device by themselves.

Secretary of State: How much would each of these devices weigh?

Director of Los Alamos National Laboratory: We don't know whether they have plutonium or enriched uranium at this point.

If it's plutonium, it's going to be [a device] a yard across, 500 to 1,000 pounds. If it's uranium for an implosion device, it's going to be 2 yards across and weigh about 8 times as much. A gun-assembled [uranium] weapon, recall that we've had an 8-inch artillery shell…, 2 feet long. And it could be as large as, say, our 2,000-pound bombs. So, it's in that range. It could be delivered by a Scud and, obviously, by a truck.

The chairman of the Joint Chiefs of Staff followed the energy secretary's assessment with recommendations relating to liaising with the Russians, alerting U.S. forces about the threat, and preparing for domestic contingencies:

JCS Chairman: It seems to me that the Russian Alpha force (which is their force that deals with this sort of problem over there) and some of our forces have the capability of getting together and determining what's gone and that we should begin cooperating with them.

We have to think about what we are going to do as far as notifying posts, camps, stations, ships, and so on about the threat. We have been going through this over the last five or six years. So we would just raise it to the level that we have practiced in the past when we have had a threat of this nature and go ahead and pass that word out. We would have to get approval from the president to do that because of the signal it sends. And you're also going to have to make a decision on the public-affairs aspect of this and how much we're going to put out and so on, so that we can deal with whatever is decided in that area.

I'll just remind you that, were there to be a problem here in the United States, the Department of the Army is the executive agent for dealing with all the other elements here in seeing that military resources are brought to bear on whatever the challenge is. The Operations Center has already been opened.

The attorney general outlined two tracks that the law enforcement community was pursuing—an FBI-led criminal investigation of the funds transfers and possible other illegal activity in the United States, and an inspection and interdiction effort spearheaded by the Coast Guard and the Customs Service:

Attorney General: One track is to exploit those narrow leads to further investigate what we have out of Detroit, increasing our electronic coverage on known Hizballah members and associates in the United States, monitoring that coverage in real time. Secondly, doing the same as to Russian organized crime to the maximum extent possible. And where the two cross, obviously that would be of interest to us. The transportation agencies (the Coast Guard and the Department of Transportation),

Customs, and the ports are all increasing their levels of activity—which has to be done. But I must say that we're looking for a needle in a haystack that way. We need more specific information so that we can direct this effort as much as possible.

The other track is to organize the domestic crisis response teams to a nuclear incident to the extent necessary. We'll seek assistance from the Department of Defense when and if that eventuality arises. We do have full coordination among the agencies. They are working well together. And the FBI has the lead in terms of looking at this as an investigation. At this point, we are not treating this as a criminal investigation but rather as a national security investigation.

Shortly after the chairman of the JCS had provided the Principals Committee with his recommendations, the DCI read from a cable additional information that had just come in from the CIA chief of station in Moscow.

COS Moscow reports that Russian liaison is convinced that local mafia has stolen nuclear material from Chelyabinsk. Quantity uncertain, but clearly exceeds what is needed for nuclear explosive devices. Insider-assisted theft occurred some weeks ago. Unimproved inventory controls did not detect it until eve of U.S. technicians' arrival to implement new security measures.

Russians now searching area and closing southern borders but have little hope of catching perpetrators. President Lebed is incensed, especially because theft was not discovered earlier and because scheduling of U.S.-assisted security upgrades at Chelyabinsk has made it impossible to handle incident quietly. High-level dismissals are likely, possibly including chief FSB. Moscow Station officer and Embassy science officer will fly to Chelyabinsk tomorrow.

Meanwhile, the Principals Committee deliberations gradually devolved into discussions of high policy and public diplomacy, with issues being raised and debated on the spur of the moment—either in the midst of, between, or after the formal briefs—instead of in the order they appeared on the meeting agenda.

On the question of whether the situation warranted the U.S. president's demanding an explanation from President Lebed, it turned out that the president had already spoken with President Lebed:

Secretary of Defense: Control, on the Lebed conversation?

Control*: The Lebed conversation was: the president called, expressed concern about our people (i.e., the Los Alamos TDYers at Chelyabinsk), said that they were reporting unusual activity, and asked what was

---

* Control is the entity composed of individuals charged with directing the course of the game.

happening. President Lebed was reassuring but provided absolutely no information whatsoever.

National Security Adviser: So, a kind of personality change.... I think that General Lebed may be tongue-tied at the moment but, if he runs true to form, he will unburden himself. He is not a devious man. What he thinks, he shows. And these are circumstances in which we have a common interest, and it will be necessary for us to establish that common interest.

On whether to increase security for U.S. forces in Saudi Arabia there was no discussion.

Regarding offers of assistance to Russia (or other nearby states) to recover the missing fissile material, there was no discussion aside from a suggestion by the DCI that was made in the context of developing a strategy for optimizing allied cooperation on gathering information about what was transpiring:

DCI: It might be useful for each of us to contact the channels from which we got this information, thank them very much, and start promoting coordination. This could be followed shortly thereafter...with another conversation telling these services that the president will be talking with President Lebed and that we have told the president that they have informed us about what has been going on. And then all of us could go immediately to work through the same types of channels to contact our allies—to get the back channels loaded before the front channels have to go...we can't have all these intelligence services getting different guidance from different governments.

On taking steps to locate and neutralize any Iranian or Hizballah nuclear capability, aside from the DCI's collection strategy, there was only limited discussion. The national security adviser himself suggested enlisting the assistance of Pakistan regarding Iran. And the secretary of state offered up an approach that built on the DCI's suggestions:

Secretary of State: It seems to me that we don't have any way yet of telling how quickly whatever the Hizballah has can be delivered to some site in the United States.

We need information very badly, and the Russians are a good source of information. To maximize their cooperation, it seems to me that our approach could be to say we are in this together, that we need to assume their good faith. It is not going to cost us anything to do that. And, when the president calls President Lebed, he needs to make clear that the first thing we need to know is how much material was diverted, when did it happen, what was it, what was the quality, was it plutonium and HEU (both HEU and plutonium are stored at Chelyabinsk) so that we have a better handle on what the threat is.

I would argue that we need to do the same thing with the Iranians. And, as you know, I have been a strong advocate of the dual-containment policy for all these years—like my predecessor—but, in this case, I think we have to challenge the Iranians to demonstrate their good faith. We need to intercept this stuff on the site in Iran rather than let it get into transit or get on a ship on its way here where it would be much harder to track. So, we need the Iranians' cooperation, I think, and the Russians' cooperation as well as other intelligence and diplomatic assets. And we should operate on the assumption that they are on our side.

There was also very little discussion regarding covert action or military, unilateral, or multilateral preemptive action. Nor was there discussion on the question of whether the United States should assist its threatened European allies.

The issue of congressional notification was addressed as an adjunct to the debate over the timing and merit of a presidential announcement to the American people:

> **Secretary of State:** One other point is that we clearly have to brief the Congress on this. And, if we brief the Congress, then there are going to be more leaks. So you'll have to be more fulsome.
>
> **Secretary of Defense:** Brief the Congress tomorrow, before the president speaks.
>
> **National Security Adviser:** Yes. You can start notifying the Congress that the president is having a leadership meeting. We will be requesting the leadership to come to the White House sometime tomorrow after the NSC meeting but well before the president's speech to the nation. And, at that time, they will presumably hear much more than is being put out publicly up until the time of the president's speech and some additional material. Do you have any further advice to us?
>
> **Congressional Liaison:** We have another initial contact tonight for the Senate majority leader. He'll be here—
>
> **National Security Adviser:** And the minority leaders.
>
> **Secretary of Defense:** One thing you might want to check with the president tonight is whether, in addition to announcing the congressional briefing in the White House before his speech tomorrow, he would also want to send one of your staffers to the Hill—one to the Speaker and one to the minority leader—in the morning.
>
> **National Security Adviser:** We will put that question to him tonight and see what his response is. I think that it is important to keep the Congress with the president on an issue as potentially devastating to the country as this. We do not need to have any kind of divergence or minimal divergence between the Congress and the president.

Although press guidance was an item on the Principals Committee meeting agenda and was key in devising a public strategy, the group backed into the issue only at the suggestion of the secretary of defense:

> **Secretary of Defense:** Another dimension to consider, Mr. Chairman, is public affairs. It is not a Pentagon dimension because we are very discreet always. But the secretary of state may have some overseas public affairs reports and, of course, the White House Office of Public Affairs. What is the story on public affairs?

In response to the White House public affairs adviser, who recommended that the White House go public as quickly as possible, the secretary of defense asked what was already known to the public, and the public affairs adviser replied:

> **Public Affairs Adviser:** What is in the public domain as of now is a dispatch from Beirut by the Palestinian News Agency, which says:
>
>> The U.S. is preparing to launch a military strike against Iran, French intelligence sources said today. The pretext the U.S. will invoke, according to these sources, is a CIA report of a planned nuclear terrorist attack against an urban target in the U.S. by Hizballah guerrillas. Both Iranian government and Hizballah spokesmen dismiss the CIA report as a new height of absurdity and said it was typical U.S. disinformation to justify an act of naked aggression against Iran.
>>
>> We have received several phone calls already from CNN and AP asking for a reaction.

Despite the number of initiatives and accompanying options still waiting to be discussed as well as the growing seriousness of the situation, fully one-third of the Principals Committee meeting was subsequently taken up with the topics of when and what to tell the American public and whether an announcement should be made at all.

The following exchange was typical of the concern expressed by some regarding notification of the public:

> **Secretary of Energy:** I'd like to make a point, sir. We have a raw intelligence report. We've had hundreds of raw intelligence reports with respect to nuclear threats for the last eight years, Mr. DCI, so why are we picking this one out?
>
> **DCI:** Excuse me. I said we had a report from the Russian intelligence service now that says some nuclear material was stolen.
>
> **Secretary of Energy:** We've gotten those reports in the past, haven't we?
>
> **DCI:** We haven't had those reports from the Russian intelligence service.
>
> **Attorney General:** I don't think there is any good to be served by creating an undue amount of public alarm where we don't have much information to go on and it may hinder our acquiring—
>
> **National Security Adviser:** You have used the word "create" public alarm. The problem is that that alarm is not being created by the U.S. government. It's going to come out of the press playing on that story on the Palestinians.
>
> **Secretary of Defense:** There had been stories in *Der Spiegel* years ago from Germany.
>
> **DCI:** Mr. Chairman, there is a difference here. One of the things that was often said publicly when those reports came out was that we had no corroboration, etc., etc. This time, we don't just have corroboration from an untested source, we have more. We saw the flap this raised in the Russian service. This service always came to us in the past, saying that they had no evidence of any of this.

Secretary of Defense: But we had corroboration that the plutonium landed in Munich in 1994.

DCI: That was not from the Russian source. The Russians tried to deny that. This time, the Russians who heretofore said "this could never happen to us; we have this stuff under control; we're terrific" came to us and said "the [bleep] have stolen something, plutonium, some stuff from us."

To me, that is the most bothersome part of all of this. All I am saying is that, if this fact comes out after we've said we just heard rumors, it is going to be a tough one to live with.

Despite attempts by the national security adviser to end the debate, it continued in a three-way exchange among the secretary of energy, the national security adviser, and the FBI director:

Secretary of Energy: If Iran was behind this, ...and they intend to detonate a device in Saudi Arabia against a U.S. facility, what will this do for U.S. prestige in the region? What will that do with respect to the geopolitics of the Persian Gulf? How will our allies react to our inability to protect ourselves, much less them? In particular, after the president of the United States has gone on TV to tell the American public that all is well and he's going to take care of it. What will that do to the political dynamics of the region, much less to the rest of the world?

National Security Adviser: Wait, wait, wait. The president of the United States is not going to say all is well and we are taking care of it. He is going to say every effort is being made. He cannot make a flat statement that we will be able to take care of it. Otherwise, the president of the United States will be thwarted by Control.

Control: That sounds like my entree, Mr. Chairman. We just received a call to the Situation Room from the Oval Office. The president is asking do we know where the bombs are and are we taking actions to recover or destroy them.

National Security Adviser: I'll talk to the president later on. That was probably a rhetorical question by the president. I think he realizes that our information on such weapons is of recent origin and that our experience in tracking Scud missiles in Iraq does not lead us to high confidence that we will be able quickly to detect the whereabouts of these weapons.

FBI Director: That question reflects considerable wisdom on the part of the president and cautions us not to help him go off half-cocked.

We are in a difficult situation right now where we know far less than we should. We have no motive. We even have the possibility that yet another group, the Chechens, may be involved. We're trying to run that down. If we announce too soon and in too large a set of decibels that there is this thing taking place and we know what it is and we are very concerned about it, we will get a high level of public concern with no guidance or instructions. Shall we be on airplanes or not on airplanes? What is happening? When is it coming? We may see sand trucks again in front of the State Department.

The FBI director helped the national security adviser in finally providing closure on the issue—at least for Day One:

FBI Director: Our job in trying to fill these blanks is going to be materially impeded as we get overwhelmed with every kind of scattershot that will come in from across the country. And I hope that whatever policy you urge on the president does not interfere with the job that's out in front of us that we must use every effort to be successful. If we are going to get there before the bomb goes off, we have to have the intelligence to do it. And we have to focus our resources on those areas in this country and abroad where we can fill those blanks. I realize that you have problems with what the public is going to know anyway, but I would certainly caution and urge you that the president take at least measured steps in what he says to the American people.

National Security Adviser: I think that's right and it's a question of careful balancing, of providing enough information to the public so that the public feels it has been adequately informed without stirring panic, and that is a very delicate balance. And we certainly will as a general policy attempt to avoid stirring panic. Whether or not that is the case ultimately depends upon the circumstances that we cannot now foresee.

In summary, on Day One, in the absence of a preapproved strategy, the discussion tended to skitter among the principals. Each individual had enormously valuable insights and inputs regarding what needed to be done, especially as it related to the discharge of the responsibilities of each. The problem was that, absent a structure, everything was ad hoc. Equally important, it was impossible to assure that everything that needed to be covered was covered. For the immediate term, for example, there was no discussion of whether the United States should submit to Hizballah demands. Longer term, there was no discussion of punitive retaliation should a nuclear attack occur.

During Day One, the participants

❏ stressed and worked toward unity of effort;

❏ decided generally to avoid the press;

❏ determined to work all possible lines of communication, even with members of terrorist groups and terrorist-sponsoring states;

❏ focused on interdiction. At no time did the principals discuss what they would do if a weapon were detonated.

National Security Adviser: In the real world, we might go on discussing this at greater length. But, in the game world, we have a terminal point and we have reached that point.

As a general observation, we are going to make every effort first to protect the United States and then more widely to protect our allies. We will seek to detect weapons movements into the United States. We will seek to deter. It should be clear—if I may speak off the record for a moment, I suspect that Control will be out to thwart us—that the president of the United States and the Congress stand together in this moment of national crisis and all of the agencies are cooperating as best they can.

Secretary of Defense: Mr. Chairman, we already are getting a good part of the president's speech for tomorrow night.

Intelligence IWG assessing incoming information.

Science/Technology IWG reviewing technical data.

*Law Enforcement IWG briefing "FBI Director" Webster and
"Attorney General" Terwilliger.*

CHAPTER 4

Playing the Game—Day Two

NATIONAL SECURITY COUNCIL
Washington, D.C. 20506
February 12, 2001

MEMORANDUM FOR

MR. EDWARD BLAKE
Assistant to the Vice President
for National Security Affairs

MR. FRANK BROWN
Executive Secretary
Central Intelligence Agency

MR. JAMES SMYTHE
Executive Secretary
Department of State

CAPTAIN B. L. KESSING
Secretary
Joint Chiefs of Staff

MS. JANICE DANFORTH
Executive Secretary
Department of Treasury

MRS. JACLYN WEST
Office of the Representative
of the United States
to the United Nations

COL MICHAEL RAWLINGS
Executive Secretary
Department of Defense

MR. FREDERICK JONES
Executive Secretary
Federal Emergency
Management Administration

MR. THOMAS JORDAN
Executive Secretary
Department of Energy

Subject: National Security Council Meeting on the Nuclear Explosion in Russia.

The President will chair an expanded National Security Council Meeting on the
Nuclear Explosion in Russia in the White House Situation Room today, February 12,
2001, from 10:00 A.M.–12:00 P.M. Attached at Tab A is the agenda for the subject
meeting. Attached at Tab B is the National Security Council Assessment that will
provide the basis for the discussion.

William B. Goode
Executive Secretary

Attachments
   Tab A   Agenda
   Tab B   National Security Council Assessment

Tab A

NATIONAL SECURITY COUNCIL MEETING ON
THE NUCLEAR EXPLOSION IN RUSSIA
DATE: February 12, 2001
LOCATION: White House Situation Room
TIME: 10:00 A.M.–12:00 P.M.

AGENDA

I. Opening Remarks   The President

II. Intelligence Update   DCI

III. Results of IWGs   DOE, FBI

IV. Policy Discussion   All Participants

    A. Crisis Management

      —Neutralize the nuclear terrorist threat to the United States and U.S. forces in Saudi Arabia

      —Assist Russia and other threatened states

      —Gain cooperation of potential transit states

    B. Contingency Planning

      —Prepare for U.S. consequence management

      —Assess adequacy of all U.S. programs to upgrade nuclear security in Russia

    C. Public Policy

      —Presidential address to the American people

TAB B

NATIONAL SECURITY COUNCIL
Washington, D.C. 20506

February 12, 2001

SUBJECT: Nuclear Explosion in Russia.

At 4:35 A.M. this morning, Washington time, the U.S. Atomic Energy Detection System detected a nuclear explosion centered approximately 40 kilometers south of Moscow. Satellite instruments estimated the yield at 10–15 kilotons. Russian officials are overwhelmed with consequence management, and cannot yet provide reliable estimates of casualties, damage, or residual radiation at this time. President Lebed telephoned the president and stated that the incident does not involve a government movement of weapons or other nuclear-related materials. Preliminary debris analysis indicates a ground burst by a uranium weapon. The U.S. has offered its full assistance once Moscow's needs become known.

Preliminary U.S. Assessment

The Department of Energy estimates that a 10–15 kiloton blast centered just outside the southernmost suburbs of Moscow could produce up to 100,000 prompt deaths, with the toll possibly tripling in six months depending upon the precise yield, residual radiation, and atmospheric conditions. These are rough estimates, and probably represent a worst-case situation. Even assuming the estimates are revised downward, however, Russia is woefully short of the medical and consequence management resources to cope with the expected casualties, public panic, and cleanup.

Chechens Claim Responsibility

At 7:00 A.M., Washington time, Chechen separatists publicly claimed credit for the detonation. They claim that the explosion outside Moscow was intended as a warning and undeniable proof of their nuclear capability. If necessary, they say they are prepared to inflict even more massive casualties. Chechnya has given the Russian government 24 hours to withdraw its forces from Chechnya and to recognize Chechen independence. Otherwise, a second nuclear explosion will devastate a Russian city, possibly Moscow.

- CIA cannot independently confirm the Chechen claims, but intelligence liaison sources report that senior officials in Moscow believe the threat is genuine.

- U.S. Embassy Moscow reports that the capital city is seized with panic. All roads out of the city are clogged. U.S. personnel are remaining in the embassy.

Hizballah Threat

Thirty minutes before the Chechen claims were announced over the separatist-controlled radio, the Iranian-sponsored Hizballah terrorist group began blanketing Western embassies in the Middle East, the international wire services, and major television stations with phone calls and facsimile copies of an ultimatum. Asserting that Hizballah too possesses nuclear arms, the sweeping and convoluted statement demanded that all non-Muslim forces immediately leave the Arabian Peninsula, that the Gulf War coalition states commit to make financial restitution "to the Muslim world…[for]…the sacrilege and damages" related to Desert Storm, and that the United States affirm that it will not intervene again in the Middle East or another Muslim state. Otherwise, there will be additional nuclear attacks like the one outside Moscow. Targets are said to include unnamed cities in the U.S., the UK, and France, as well as U.S. forces in Saudi Arabia.

- Although the ultimatum denied responsibility for the Moscow detonation, and it expressed sympathy for the Russian people, it insisted that the explosion proved that nuclear weapons were now readily available and that the "use threshold" has been crossed. It claimed that Hizballah has acquired them.

- CIA and State Department offices in Tehran and Beirut (note: Hizballah maintains several camps in Lebanon's Bekaa Valley, which is the most likely site for any terrorist-run nuclear-weapons facility) have been tasked to report immediately on Hizballah activities, on the group's technical capability to assemble nuclear weapons, and on how government officials in Tehran are reacting to the group's claims. No response has been received yet.

- The South African Intelligence Service has told the CIA station chief that Pelindaba Technologies Ltd. early last month airfreighted three sets of nuclear triggers, wiring harnesses, and conventional high explosives to a commercial entity in Dubai. CIA is attempting to track the subsequent movements of this shipment.

Iranian "Spy" Apprehended in New York. Alerted by the February 10th CIA report that U.S. ports were being targeted, and due to this morning's heightened security in response to the explosion in Russia, harbor police in New York City just two hours ago apprehended an Iranian national attempting to sneak out of a controlled area of the waterfront. The suspect is a student enrolled at Emory University in Atlanta. In his parked car were several harbor sketches and a laptop computer. Although the suspect insists he is innocent, FBI agents just arrived at the scene report that the computer memory contains copies of previously sent Internet messages with reports and sketches of four U.S. harbors: Charleston, Norfolk, Baltimore, and Philadelphia. The e-mail address on the messages belongs to a private company in Tehran.

- The FBI Director has activated the Federal Emergency Response Plan (FERP) and is establishing an interagency command post to facilitate counter-nuclear terrorist measures and disaster planning. The Department of Energy has alerted the Nuclear Emergency Search Team (NEST), which is dispatching technical experts to the FBI Field Office and co-located FERP Command Post in New York City.

Stolen Russian Fissile Material

Although details of the device detonated near Moscow remain scant, the Department of Energy advises that a 10–15 kiloton yield would be consistent, for example, with a simple gun-assembled weapon that might use 60 kilograms of HEU in finished parts. Allowing for losses in casting and machining the bomb parts, such a device might consume 70–90 kilograms of the 200 kilograms of weapons-grade HEU that Russian authorities have determined was stolen from Russia's Chelyabinsk-65 in late December. Thus, enough HEU remains unaccounted for to produce a second weapon of approximately the same yield. In addition, the 20 kilograms of plutonium reportedly taken from Chelyabinsk-65 in the same theft would suffice to construct three weapons of approximately 10–15-kiloton yield. Thus, enough fissile material remains at large to produce one more uranium and three plutonium weapons.

CIA, FBI, and U.S. National Laboratory liaison contacts in Russia all report that the fissile material was stolen by a Chelyabinsk employee (since disappeared) and removed by local criminals (a Chechen gang, also now disappeared). Moscow has provided no information on the ultimate buyers but clearly is focusing on the Chechen separatists. The United States has informed President Lebed, the NATO Allies, Israel, and Saudi Arabia that we have credible reports that Hizballah—possibly with Iranian assistance—acquired at least some of the stolen material and has managed to assemble one or more nuclear explosive devices that Iranian and Hizballah officials believe to be reliable. We have also informed the Russians and the

Allies that separate sources indicate the Iranians or Hizballah purchased three identical sets of nuclear triggers and other components of an implosion device. U.S. analysts believe it to be highly unlikely that Hizballah or Iran was involved in the nuclear detonation outside Moscow. Thus, the Chechen separatists, Hizballah terrorists, and possibly Iran may now possess nuclear weapons.

Parameters of a Hizballah Nuclear Attack

Department of Energy nuclear-weapons experts believe it is likely—although not certain—that any Hizballah nuclear attack on a U.S. target would utilize an improvised plutonium implosion weapon. DOE analysts assume that Chechen and Iranian/Hizballah purchasers would choose to minimize complexity by buying HEU or plutonium rather than acquire some of each. They also believe that Iran or Hizballah has purchased components of an implosion device. Assuming that the Iranians or Hizballah build three weapons with the 3 sets of nuclear triggers, the most likely yield would be approximately 10–15 kilotons each, approximately the same as that of the blast this morning outside Moscow.

- Assuming no prior evacuation, a full-yield detonation of such a weapon within any of the metropolitan areas visited by the Iranian suspect apprehended in New York could produce about 100,000 prompt deaths.

-4-

Day Two of the exercise was the crisis management phase. It began with the executive secretary's announcement of the NSC meeting and an accompanying intelligence update on the nuclear detonation outside Moscow. All participants received the NSC staff summary of the situation, covering all known facts of the nuclear blast, threats and other public statements by Hizballah and the Chechen separatists, and details of the nuclear theft that were provided by the cooperative Russian government. The day consisted of two more rounds of interagency meetings.

In Round Three, the Intelligence, Law Enforcement, and Science/Technology IWGs met again. They evaluated the changed situation, prepared recommendations, and briefed their principal officers (as they had on Day One).

The nuclear detonation in Russia galvanized the IWGs into action:

❒ The Law Enforcement IWG continued its efficient division of labor: the FBI launched a criminal investigation of the apprehended Iranian student, and the Coast Guard crafted a layered defense against a seaborne attack.

❒ The Intelligence IWG prioritized the threats to the United States and considered a list of possible covert actions (none of which was later considered at the NSC meeting).

❒ The Science/Technology IWG refined its estimates of Hizballah's capabilities and the number of nuclear weapons that might be loose while FEMA outlined initial (but already tardy) steps in consequence management.

The exercise culminated in Round Four, with the president chairing the expanded meeting of the NSC to manage the crisis.

The NSC was as frantic as the IWGs. Ideas were advanced to invade the Bekaa Valley, to contact Iran, and to help Russia; but there was no clear focus. A superpower with great intelligence and military capabilities and economic power was humbled by a nuclear threat. Time was against all the grand schemes advanced.

The president began with a scene-setter that stressed that a nuclear weapon had been detonated in anger for the first time since 1945:

> **The President:** This will be actually the first National Security Council meeting since the inauguration. And it is on a weighty issue.
>
> For the first time since Nagasaki, we have seen a nuclear weapon exploded in anger. It has occurred in Moscow—unquestionably using materials that were obtained within the Russian system.
>
> Unfortunately, there is no certainty that those materials are still confined within the Russian Federation. And as a consequence we have much to deliberate because we have heard these threats from Hizballah.
>
> And all of us will have to work together and see to it that the agencies are coordinated in their response to what is a serious threat of a sort that the American people have not faced in the past.

After the president's opening remarks, the DCI was asked to provide his intelligence update. In the interest of time, he did not repeat what information had already been disseminated; he focused, instead, on two key points that required policy discussion: one was the extent to which the Iranian government was supporting Hizballah in the effort and the other was the issue of movement. (The DCI advocated concentrating search efforts in the Middle East, not realizing that the device had already left the region.) Regarding what diplomatic or intelligence sources were saying about the deportment of Iranian officials and the difficulties in getting additional intelligence on terrorist activities, he had the following exchange with the president:

> **DCI:** The weight of the evidence would lead us to concede that it is not a joint effort. We don't rule out the possibility that the government is somehow complicit in this…. The Iranian government wants the nuclear materials for itself and not for the Hizballah. And all the evidence we have points to the fact that it was actually stolen from the Iranian shipment…. One of the advantages, had the Iranians been complicit, is that we have better penetration of the Iranian government than we do Hizballah. The disadvantages of the Iranians not being involved is that the Hizballah is tough.

> **The President:** Are you still following the policy that we do not have any contacts with those who have violated human rights that has precluded our dealing with Hizballah?

> **DCI:** Sir, I am not allowed to address that in front of— We do have a source. While he is relatively new and untested, what he has told us so far has been borne out. His description of the Iranian officials' reactions— unless he's a double agent and he's passing disinformation—is that they are very concerned. Indeed, that is both good and bad news. Their concern is that they see this as creating major problems for their own nuclear capability at a minimum as well as the rest of their policy. On the other hand, had they been successful, we would not have liked that either.

After the intelligence update, the president called on the secretary of energy, who was aided by the director of Lawrence Livermore National Laboratory, for a technical assessment.

The secretary reported that, given that Hizballah possessed 20 kg. of plutonium, the most credible threat Hizballah poses to the continental United States is a radioactive dispersal device composed of the plutonium and a high explosive to disperse it over a large area. He cautioned, however, that if Hizballah is aided by Iranian technicians or others the assessment then would be that Hizballah has the capability to fabricate three plutonium implosion devices:

> **Secretary of Energy:** They could have three devices as we speak because they have had the material now for a month and a half. And, given the technical assistance, there is no question that they could put something together that would have a low yield.
>
> This is going to change the dimension from both a political and practical standpoint in terms of the threat to the continental United States. A dispersal device obviously is not going to cause anywhere near the damage and casualties that a detonation will. So probably the single most critical intelligence requirement is to determine who is behind this and who is working with these people to either fabricate a device or put together a dispersal device.
>
> With respect to the assets we have, sir. We have upwards of 200

technical scientists, engineers, and others who are part of NEST. All are available and have been deployed. We are working with the FBI, who is leading the effort in the United States out of New York City. We have sufficient capability to deploy to the ports on the eastern coast. The problem, as you know, sir, given the limitations of detection with current technology, is that it is fairly man intensive. So, if we have to search an entire port or a whole series of cargo ships, it is a long and drawn out process. Obviously, the better the information we have as to location the more likely we are going to be able to locate one of these devices.

In response to the president's comment that the United States needed to get as much information as it could regarding the composition of the Iranian program (How far had Iran progressed? Had Iran acquired the components that would lead to a near-term plutonium implosion device?), the secretary of state said that determining this takes the diplomatic challenge to a dramatically higher level:

Secretary of State: It is one thing for them to cooperate with us in identifying, locating, and destroying the Hizballah weapon. It is quite a different thing for them to come clean about their nuclear-weapons program, which we know they have been pursuing for many years. And they are not about to give that up lightly. In fact, they may in the current context feel they themselves are in some danger because of our doubts about their possible involvement.

I think my recommendation, notwithstanding the technical issues, would be that we have to find these devices and that we ought to place less of a priority—despite the importance of the information—on identifying specifically what kind of device it is. Nor would I want us to sacrifice our ability to neutralize these devices for the sake of getting more information about the Iranian program, which may be unattainable under any circumstances.

The attorney general then weighed in, urging a layered defense beginning with a picket line 200 miles from U.S. shores:

Attorney General: Mr. President, from the law enforcement side on the detection issue, we can do a layered defense reaching out as far as possible with as much detection equipment as we can lay our hands on (if we can get DoD assistance for both transport and personnel transport) and try to do our part to help effectuate this containment policy by physical surveillance.

We do think, however, that it is prudent and necessary until we know more to set up a picket line along the east coast, particularly to intercept small shipping that is not required to provide 24-hour notice to the Coast Guard and Customs of landings in U.S. ports. And to back that up with an intensive effort in the ports themselves on the off chance—and I frankly think personally it is a very slight chance—that there is a device already here.

However, there is a political dimension. We are likely going to throw a wrench into the gears of commerce that will almost shut down commerce on the East Coast in the major ports. And it's likely to require a great deal of cooperation from foreign governments, both in terms of ship boardings and preembarkation searches of both aircraft and commercial cargo ships. We can do that with proper DoD support. I think that the secretary of defense and I can conclude that there is a national emergency that renders use of military assets appropriate.

The attorney general also warned that the Iranian student might have been only one of several such spies, and thus there might be other targets about which the United States is not aware:

> **Attorney General:** Based on the information obtained as a result of the apprehension of this Iranian student, we have set up joint operations centers for contingency purposes in New York and the other four port cities that were mentioned. The basic purpose of these joint command centers is to have crisis management teams standing by in place should we have a hit on one of the detection efforts or better intelligence about where the device is located.
>
> On the Iranian student arrest, it is very difficult to know what to make out of that information. It is obviously highly suspicious on its face. But the FBI is pursuing a full field investigation both in Atlanta and in New York on the individual to determine any connections to any Iranian government or Hizballah assets. We just do not know much about this person at this point.
>
> There is also a risk in putting too much weight on that particular information. We do not know whether this is one student that looked at five ports, or is one of five other students that also looked at five other ports. So we have to be careful not to assume too much at this point from that information.

The president then asked whether the size of the weapon precluded air delivery and, if it did not, stated that it would be desirable to have inspections of flights coming into the United States from the Middle East. The director of the FBI answered that the inspections were being done not at embarkation points but at airports in the United States. The secretary of defense argued that focusing on our own airports was bringing the inspections too close to some of our large cities; the attorney general agreed, requesting a policy decision for overseas inspections as well as Department of Defense support.

This exchange in turn triggered a lengthy discussion, led by the secretary of defense, on a broader recommendation:

> **Secretary of Defense:** This is a point in history—as you have said at the beginning of our meeting—that is extraordinary. And I think it gives us an opportunity for bolder action where you would start a new era hopefully more peaceful than the 50-year period started by President Truman's bold actions.
>
> If these perpetrators get away (and we now have an explosion near Moscow so it is not like yesterday—it is not an "if"), our world order will collapse and we will have repeated actions of clandestine delivery. So we have to make every effort not to just try to minimize further destruction in the next few weeks but to minimize the damage to our world order and world peace.

> The chairman of the Joint Chiefs of Staff and I and our staff in defense have thought about this through the night and this morning, and we have come down to three ways of handling this situation that we have been working for the last 50 years. They are defense, deterrence and dissuasion, and counterforce to disarm.

Regarding defense, the secretary of defense recommended closing the borders for 48 hours and letting through only official U.S. and allied aircraft, which would, it was hoped, allow time to install an inspection system abroad, deploy more personnel, and acquire more information on the entire situation.

Moving on to deterrence and dissuasion and response, the secretary of defense offered the following:

> **Secretary of Defense:** I think it is your intention, Mr. President, to give a talk to the nation and to the world this afternoon. This is the time to bring deterrence to work.
>
> We would recommend that you would make clear that any group or entity that has acquired nuclear materials contrary to the Non-Proliferation Treaty must within 48 hours inform the UN Security Council of the exact location where this material is held and make it available for immediate retrieval to render it harmless. Failing that, such a nation or group or entity would be subject to punitive action with all the means at our disposal.
>
> We would recommend—for the secretary of state's evaluation, of course—that you do this as a Big Five action with the United States leading, pending urgent consultation and coordination with the Big Five. You would try to bring them on board but not be slowed down by them. And then you could say in the speech that, given our intelligence brief this morning, we enjoy very close cooperation with the Russian government and are looking forward to assistance from Iran, to bring them over to our side. You also have to deal in the speech with the momentous event that is already in the news this morning—the detonation south of Moscow. That is a case in point. The perpetrators must not be left free, unpunished. And you might state that, in the event a weapon is used—that also helps in deterrence—the entity responsible will be eliminated with no sanctuary anywhere and all measures necessary.

In response to the president's request for comments by the others regarding the suggestions, the U.S. ambassador to the UN offered a fourth option: that the United States try buying the material from someone within Hizballah—an option earlier considered and implemented with respect to Ukraine and North Korea. The secretary of defense agreed, saying that we wanted to use all means that were helpful whether or not they left a bad taste and referring to the utility of using the Atomic Energy Act of the 1950s and the terrorism information reward legislation of the 1980s as bases for such an initiative.

Regarding all of this, the secretary of state cautioned against making too broad a declaration. He suggested the U.S. demand be made in such a way that it excluded the Iranian government's illegal possession of nuclear materials (despite its being a signatory to the Non-Proliferation Treaty) and focused instead on illegal possession by a private group:

> **Secretary of State:** I think, if we mixed the Iranian nuclear program with the problem of the current threat, the problem of terrorism, that we are less likely to solve the immediate problem. So I think we need to be very carefully focused. And I think that can be done by refining the language of your proposal, Mr. Secretary, to make clear that we are talking about entities that have made explicit threats against the United States and our allies.

Generally, everyone agreed regarding the advisability of using all the resources available in conjunction with DoD to push the inspections as far back as possible to the embarkation point.

Returning to the question of shutting down the borders, the president agreed with concerns voiced by the attorney general and noted that there would be hardships for people in Canada and Mexico:

> **The President:** It is not clear that we have to shut down the borders in the sense that those who are traveling by foot cannot come into the country. As long as noninspected automobiles do not come in, there should be no danger.... I do not know what the state is of our radar capabilities along the two borders, but some years ago they were not awe inspiring.
>
> **Chairman of the JCS:** We have the ability to put up AWACs and so on at the present time to augment them, Mr. President. We do not do so on a normal basis because of the cost. For two days, we could do that without much difficulty.
>
> **The President:** Please do. Because the best way of sneaking a weapon into the United States is by a short aircraft flight from Canada, Mexico, or the Bahamas. Or by just a helicopter for a very short flight, thereby getting past all of the Customs and other officials at the border. So we must do what we can with regard to the detection of aircraft movements.
>
> It is more important for us to be able to interdict with some degree of success such aircraft or automobile movements—particularly truck movement—than it is to prevent people who for one reason or another have to come back into the United States. So, I hope we can organize a working group that will more precisely define what we mean by shutting our borders.

The discussion veered dramatically after the secretary of energy raised the issue of (Syrian and Iraqi) state support to terrorist groups, such as Syria's support of Hizballah in the Bekaa Valley, and the need perhaps to get more serious with respect to who was really responsible for the current situation. After his remarks, there commenced a lengthy discussion of defense and diplomatic efforts to bring pressure to bear on the Syrians, to root out the Hizballah in the Bekaa Valley, to protect U.S. forces in the Middle East (e.g., using air defense systems to protect against Scuds), and so forth.

Only after the secretary of state cautioned the group did the discussion gravitate—albeit temporarily—to the critical matters of the moment:

> **Secretary of State:** What is going to be the impact of that activity on our ability to find the weapons or on Hizballah's ability to conceal them or

move them were their intentions to do so? It seems to me we have to keep our eye on the specific objective here.

The President: Well, we have a short-range objective, but we also have the long-range objective of making a point that will be remembered for the decades ahead by anybody who would like to emulate either the Chechens or possibly Hizballah. So we have to balance those two objectives. You are quite right, Mr. Secretary, to point to continued attention to the short-term objective.

Secretary of State: There is also that—after the American people are no longer in danger.

The attorney general followed up, saying that a policy decision also needed to be made about U.S. planning for the consequences should intelligence and detection fail, a device turn up on U.S. soil, and an explosion or dispersal of some sort occur. Noting that FEMA would step in should a disaster occur, the attorney general warned that there still remained the problem of resource allocation should the Russians request assistance:

Attorney General: We cannot afford to spread ourselves too thin in sending our resources abroad when we may still need them near term at home.

The President: I think quite clearly the priority for our resources will continue to be here at home, that anything that is not required here could possibly be sent overseas to help the Russians. And at least some symbolic help is desirable under these circumstances. But not so much help that it begins to cripple us.

The UN ambassador said the Europeans were also concerned about a detonation on their soil, adding that the Europeans do not accept at face value the Russian assessment of the amount of HEU and plutonium missing and that the ambassador herself was surprised that the intelligence community so readily accepted the amounts cited, given Russia's poor accountability procedures—even in 2001. The DCI noted by way of explanation that the community accepted that that is what the Russians believe and took into account questions regarding the Russians' competence.

Joining the discussion about the Europeans, the chairman of the JCS summarized DoD consequence-management arrangements there, noting that the U.S. commander in chief in Europe had been working with NATO elements to establish a joint task force that should minimize drawing down on U.S. assets.

At this point, the vice president, who had been quiet throughout the NSC meeting, entered the discussion with a caution to the group:

The Vice President: I am concerned here that the clock is ticking. That we are talking about a range of issues, some peripheral, some more essential, while we have no reliable information to suggest that a device might not be steaming into one of our ports at this moment.

May I recommend to you and to this council that we concern ourselves with the immediate containment effects in invoking and taking whatever decisions are necessary to protect what might be hundreds of thousands of American citizens before concerning ourselves with other matters.

It seems to me that there are a number of uncertainties and technical issues that need to be resolved here in order to make such a decision and that it behooves us to focus on that order of priority.

The vice president's remarks gave an opening to the FEMA director to speak:

> **FEMA Director:** Mr. President, I appreciate being allowed to continue as a participant in a cabinet meeting in your administration. The vice president's point is well made. We have heard two rationales this morning for some kind of annunciation by you of a national emergency, one to exercise authority to close the borders and the other for you to have a podium to speak to the international community.
>
> I think a third important consideration is for an emergency declaration that would empower us to use all the resources available for consequence management and uninhibited access to disaster funding. This would let all the participants in the federal community know they had unconstrained access to resources in addressing this national priority. It would be a very important and helpful decision and would also be seen by the public as a positive, hands-on response to this threat.

The president agreed on the need to examine the consequences of a possible detonation and stated that would be done. He went on, however, saying that now that there was a general plan for what must be done it was time to turn to the issue of further congressional consultation:

> **Congressional Liaison:** A couple of points. The first is that House Speaker Cassis, as you might know, is quoted in the papers as saying that he is disappointed that the Congress has not been consulted or kept apprised of the situation. But he feels it is in the best interest of the country to support the president in this matter at least at this point.
>
> Also, as you might know, last night on *Larry King Live,* Senator John McCain sharply criticized the administration for what he called the troubling secrecy, equivocation, and lack of strong leadership on an issue of utmost urgency to both the Congress and the American people. He also questioned why a state of emergency has not been declared and called for full disclosure of the gravity of the situation and what this administration plans to do about it.
>
> Finally, we have been flooded with calls from members demanding a full updated briefing on the situation. Senator D'Amato from New York called to demand an explanation for the stepped-up FBI activity in New York and whether or not it was related to the recent arrest of the Iranian student.

The congressional liaison then recommended a briefing with the leadership as rapidly as possible, updating the situation and letting the leadership take it from there as far as briefing the various members of Congress. The president agreed.

Commenting rather tartly that, thanks to Control, he had lost 48 hours in dealing with this issue, the president turned next to his public affairs adviser, saying that he hoped he had not been idle during that period of time:

> **Public Affairs Adviser:** No, Mr. President. In fact I think you have caught bits and pieces of all the morning news shows. They have been totally preempted by live feeds of thousands of refugees fleeing Moscow, interspersed with many talking heads, including people like Larry Collins (author of *The Fifth Horsemen*, a book about a nuclear terrorist incident in Manhattan in the late seventies) describing the horrors of a nuclear explosion in Manhattan. We have had Pierre Salinger on NBC, who claimed that the Russian explosion was an accident, that it was not a terrorist-detonated explosion; but the Chechens, of course, are taking credit for it.
> The *New York Post* afternoon edition—
>
> **The President:** Pierre Salinger was one of your predecessors, as I recall.
>
> **Public Affairs Adviser:** —which is on the Internet now, which hit the streets of Manhattan at 10 o'clock, says as its lead paragraph:
>
>> Israeli intelligence sources said today that six Russian tactical nuclear weapons had been stolen and are believed to be in the hands of Iran, Iraq, Libya, and the Hizballah terrorist organization. An Israeli government spokesman said the U.S. administration had informed Israel that unconfirmed reports indicated unspecified nuclear materials had indeed been stolen from a Russian facility. Israel is now calling out reservists, and a massive buildup of tanks and APCs is under way on the border with Lebanon.
>
> A military commentator of the newspaper *Ha'aretz* was on CNN speculating that the IDF was planning to roll over the entire Bekaa Valley with a scorched-earth policy.
> A few minutes before this meeting began, Mr. President, there was an AP news alert bulletin, which said that Middle Eastern terrorist organizations are attempting to smuggle a stolen Russian nuclear weapon into a major U.S. port:
>
>> Administration sources said today an emergency NSC meeting has been called by the president for 10:00 A.M. EST. These same sources said the president is expected to order a maximum alert to the FBI, FEMA, Customs, and all state and local law enforcement agencies.

The reference to the reported Israeli activities in the Bekaa Valley triggered another exchange on projecting U.S. military force into that region—in cooperation with the Israelis. Discussion of this was halted—albeit briefly—when the public affairs adviser told the president that he had neglected to mention that, according to an Associated Press bulletin, traffic coming into Manhattan on the Triborough Bridge was running at only about 10 percent of normal volume.

The attorney general again pressed for the need to focus efforts on trying to find the device and successfully rendering it inert—notwithstanding the import of the longer-term foreign policy issues. Not to do so, he warned—not only as the attorney general but also as the president's former campaign manager—would result in the president being a "dead duck" politically. Despite this warning, the discussion again veered back to the Middle East, the ramifications for a possible Arab-Israeli war, Israeli cooperation in the Bekaa Valley, and so forth, until the president again turned to his public affairs adviser, who noted that he now had Associated Press follow-up on Manhattan's outgoing traffic: it was running 90 percent above normal volume.

Turning then to his FEMA director, the president asked how prepared they were to deal with the consequences of a hypothetical detonation in the United States, noting that this goes beyond the Mississippi overflowing its banks from time to time:

> **FEMA Director:** I have to caution you that, despite all our efforts during the last four years of the previous administration to enhance civil capability, we are still very dependent on the military for lift, engineering resources, and certainly medical resources. So, as we look at the prioritization of those assets, we have a very strong claim on some of those if something should happen in one of the major metropolitan areas.
>
> We will clearly have a major exodus that we will not be able to control. We will work with public affairs and look to your pronouncements to show that we are on top of the problem. But clearly we will be dealing with a massive panic in whatever city that might even think they are going to be a target.
>
> We have got some critical assets from all of our departments and agencies that, if we get a declaration, we can begin to pre-position to improve response times. These include our Metropolitan Medical Strike Teams (we have those in 20-plus metropolitan areas), some urban search and rescue capabilities, some DMORT [Disaster Mortuary Response Teams] options. The real concern, though, is working with the congressional delegations and the state governors to ensure that their plans are in place and they can begin immediately to take precautionary measures. Here we are in February, so, if something occurs in these northern port areas, there will be a lot of problems with sheltering and care of evacuees.
>
> We are going to incur casualties. We are going to have to deal with the consequences of that. But, in terms of supplies and critical procedures, they are in place. We are working with the director of military support [DOMS] very closely, and we can have continuing advisory information for you.
>
> We have an inventory of the national critical infrastructure that is at risk and a national inventory of all the federal resources available to state and local authorities—based on 1997 Nunn-Lugar provisions. So we can give you a pretty good indication on what your resources are. But we are certainly going to have to be in a damage-control mode, and we are going to take casualties.

After a number of questions relating to the implications of radiological dispersal vs. an explosion and predetonation and postdetonation options, the president inquired about the medical aspects of the crisis. Noting that they work hand in hand with FEMA, the director of the Public Health Service said that the emergency services are ready to go as soon as the president issues his emergency declaration.

Having ascertained that DoD was ready to assume domestic transport, engineering, and other responsibilities, and alluding to the need to have a plan in hand to deal with cleanup operations, the president concluded the meeting with a discussion of potential numbers of casualties—with Control noting they could run as high as 100,000 deaths.

The president closed with a firm directive to inform the American people and evacuate cities should that prove advisable.

In the end—

> *Nothing was decided or even suggested that would have prevented a nondescript freighter with a crude nuclear device in its hold from arriving in Baltimore harbor in two days time.*

"President" Schlesinger leading NSC discussion as "Vice President" Peterson and "Secretary of State" Holum look on.

"Chairman" of the JCS Meyer and "Secretary of Defense" Iklé conferring.

"Congressional Liaison" Bopp and "Public Affairs Adviser" de Borchgrave monitoring developments.

CHAPTER 5

Assessing The Game— A Few Observations

Wild Atom—How Realistic?

Although approximately 70 well-qualified players and controllers put their best efforts into Wild Atom, an artificial exercise never quite replicates what officials would do in a real-world crisis situation. The scenario was realistic, but constraints intruded. Time was short. Participants operated without benefit of staffs, reference materials, or the context of daily events that would have helped prepare them for such a crisis. Crucially, the discussion was unclassified. The authors of this report have made every effort to highlight conclusions and recommendations from the exercise that they believe accurately reflect the realities of the nuclear terrorism threat and current U.S. policy and capabilities.

Nuclear Terrorism Threat

Just a few years ago, many U.S. counterterrorism experts did not believe that terrorists would use weapons of mass destruction. The experts argued that terrorists considered conventional explosives to be sufficient for their purposes. Moreover, the experts believed, groups would not use nuclear, biological, or chemical weapons for fear that such abhorrent tactics would jeopardize their state sponsorship or popular support.

That judgment changed in March 1995 when the Aum Shinrikyo cult released sarin nerve agent in Tokyo's subway, killing 12 people and causing 5,500 to require medical treatment. The subsequent investigation revealed that the cult had developed and tested both chemical and biological weapons (on a purchased Australian ranch), had investigated uranium enrichment, and even had a member of the group working inside Russia's Kurchatov nuclear research facility. But Aum Shinrikyo may not have been unique. At sentencing for the February 1993 World Trade Center bombing, the judge concluded from the evidence that the bombers probably had packaged sodium cyanide with their explosives. He theorized that the blast had vaporized instead of spread the lethal toxic agent.

Some complacent experts still maintain that terrorists lack the capability to build nuclear weapons, but this belief ignores the facts. Nuclear-weapon technology is now 52 years old. In the aftermath of the Soviet Union's collapse, weapons-usable uranium and plutonium have been stolen and transported illegally across international borders although the materials seized to date have been in quantities insufficient to produce a nuclear explosion.

Although Russian atomic energy minister Viktor Mikhailov insisted that no nuclear weapons had gotten loose, Russian presidential contender General Aleksandr Lebed implied that they are vulnerable to theft; he warned in November 1996 that, if Russian nuclear weapons were not properly protected, it could lead to terrorism of an unprecedented scale. More recently, in September 1997, Lebed asserted that as many as 100 suitcase-size Russian nuclear weapons had disappeared and might be sold to rogue nations or terrorists.

Meanwhile, Russian nuclear materials and the know-how for building a bomb are not as well protected as Russian weapons. Several European governments have seized smuggled nuclear material of probable Russian origin. Moreover, despite U.S.- funded efforts to find gainful employment for former Soviet nuclear-weapon specialists, many remain economically distressed. In October 1996, the director of Chelyabinsk-70, one of Russia's two nuclear-weapon labs, committed suicide. By all accounts, he had despaired that the institute's scientific staff had not been paid in months.

Perception and Response

Wild Atom participants were sobered by the extreme risks and uncertainties of the nuclear terrorism threat that they faced. They tended to avoid the unthinkable consequences of failure to prevent an attack and were frustrated by the slim odds of a successful interdiction. Some did not quite know how to respond. Although many possessed expert knowledge of proliferation and nuclear matters or were specialists in counterterrorism, their reactions suggested that the exercise raised their awareness about the potential for nuclear terrorism and the shortcomings of U.S. defenses. At the end of the two-day exercise, all agreed that the scenario was realistic and that the United States is vulnerable.

Among several tendencies observed, especially in the two cabinet-level policy meetings, two stood out as particularly disruptive: (1) focusing on traditional, agency-specific approaches and responses and (2) digressing to longer-term goals.

❑ *Traditionalism.* Participants gradually divided into two loose groups, each focusing on its narrow agenda and failing to integrate policy options with the other: the foreign policy, defense, and intelligence group concentrated on projecting power into the Middle East, whereas the domestic and law enforcement agencies concentrated on search activities within the United States. Although these polarized groups never crystallized into opposing camps, the dichotomy impeded development of a coherent, multitrack strategy. What emerged instead was an assortment of independent and sometimes conflicting initiatives. Discussants vacillated, the national security adviser had difficulty aligning the work of the two groups, and the principals had difficulty achieving closure. At the end of the day, there was substantially less than a comprehensive understanding of the problem or a strategy for dealing with it. Indeed, only gradually did participants reach consensus that the overwhelming imperative must be to intercept the device well away from U.S. shores.

Observers of Wild Atom concluded that dealing effectively with today's threats, which increasingly have both foreign and domestic components, will require integrating the capabilities of the national security establishment with those of the domestic agencies. No overarching authority short of the president (and the rarely assembled cabinet) currently exists to oversee such combined efforts. We believe that the traditional concept of the national command authority may be outdated and should be reviewed.

❑ *Digression.* Participants frequently lost their focus on the immediate problem and digressed on longer-term goals. For example, some saw an opportunity in Iran's initial responsibility for purchasing the plutonium (that Hizballah later stole) to coerce Tehran into giving up its own nuclear-weapons program. Others saw a potential opening to draw Syria into cooperating in a search of the Bekaa Valley to eliminate terrorist camps there. Were it not for repeated exhortations from the secretary of state and the attorney general, further digression would have occurred with valuable time lost. Understandable as these excursions were, they distracted other participants and impeded movement toward a coherent action plan to deal with the immediate nuclear threat.

Among the insights and recommendations that emerged, the most important lessons derived from misperceptions and false assumptions. Even discounting for the nature of a training exercise, the following observations of Wild Atom could well occur in a real crisis:

❑ *There was a tendency to take the available intelligence at face value.* Few participants—with the exception of the Intelligence IWG that wished for reliable human intelligence (HUMINT) to resolve all ambiguity— questioned the accuracy or completeness of the intelligence they received. In more than five hours of deliberations by the principals, there were only two brief references to the possibility of deception, and no one pursued either comment. On Day Two, moreover, most participants wanted to focus U.S. efforts on the five sites that the arrested Iranian student had surveilled even though the attorney general noted: "We do not know whether this is one student that looked at five ports, or one of five students who each looked at five ports."

❑ *Treating the problem as though it began when the players first learned of it led to false assumptions and responses that were too little, too late.* Most participants initially assumed that Hizballah began trying to build a nuclear device when it acquired the stolen fissile material instead of many years earlier. They implicitly assumed the theft had occurred shortly before Russian authorities detected it instead of nearly two months before. Some of them worried about only one bomb, whereas enough material had been stolen to manufacture five. And the DCI, assuming that too little time had elapsed for the terrorists to have assembled and moved a device, repeatedly urged efforts to "contain" the device in the Middle East (whereas, unknown to him, the terrorists had already placed it aboard a freighter bound for Baltimore). Thus participants generally underestimated the threat, overestimated the time available to counter it, and focused their countermeasures in the wrong region.

❑ *Insufficiently probing the perpetrators' motives and cohesion left potentially rewarding policy options unexplored.* The Hizballah faction's willingness to use nuclear weapons might well have alienated some members, but only one principal suggested trying to buy back the material, and no one suggested an effort to encourage disaffection in the terrorist ranks and to try to recruit a defector with vital knowledge. One working group wanted to orchestrate an information warfare campaign to dissuade Hizballah from attacking, but the idea was never raised in the cabinet-level sessions.

❑ *One-problem-at-a-time blinders caused participants to let valuable time slip away and to neglect other concerns.* First, by concentrating on the known theft, participants largely ignored the possibility of other lapses in Russian security and possible multiple seizures. (The Science/Technology IWG did raise this concern, but it was never discussed at the cabinet-level policy meetings.) Second, virtually the only actions to come out of the first day were efforts to gather more information. Possible military, covert, or defensive actions were delayed. Third, the principals were so focused on searching for the device that they did not discuss until late on the second day how to cope with the aftermath of a nuclear detonation in the United States.

❑ *Old thinking got in the way. The working groups tended to revert to familiar but inappropriate analogies.* They looked to drug interdiction measures despite the fact that a nuclear device must be intercepted well away from U.S. shores. They focused on a missile attack despite the obvious fact that such a means would be the least likely for terrorists (or even Iran) to use against the United States. When they did discuss the more likely possibility that terrorists would use improvised means such as commercial aircraft or the cargo vessel called for in the Wild Atom scenario, some participants still concentrated on Customs Service searches at U.S. ports and airfields. Like some other responses, these seemed to be examples of officials doing what they have been conditioned to do instead of what the situation requires.

❑ *The deliberations were not sufficiently action oriented.* In some working-level interagency sessions, players discussed the situation too long and rarely came to closure on suggestions for policy or action. Worse, the principals failed to transmit most concrete proposals from the working groups to the decision-making senior meetings. And officers in the charged and highly political atmosphere of the senior Principals Committee and NSC sessions acted deferentially and spoke too often in generalities.

Bucking these trends, the DCI, given the extraordinary circumstances, persuasively urged broad sharing of the available intelligence and waived concerns about source protection. He also proposed forceful measures to shut down transportation out of the Middle Eastern area (not knowing the nuclear bomb was no longer there).

Intelligence

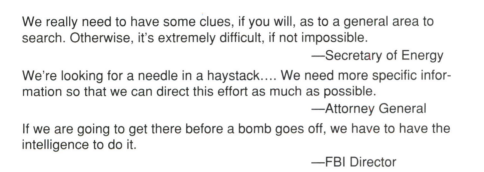

We really need to have some clues, if you will, as to a general area to search. Otherwise, it's extremely difficult, if not impossible.
—Secretary of Energy

We're looking for a needle in a haystack…. We need more specific information so that we can direct this effort as much as possible.
—Attorney General

If we are going to get there before a bomb goes off, we have to have the intelligence to do it.
—FBI Director

The intelligence provided to participants in Wild Atom was deliberately incomplete and subject to different initial interpretations. Yet all agreed that the information provided was probably the most that would be known were a real nuclear terrorist incident unfolding. In the real world, there would have been more background noise of contradictory information and other distracting events.

Participants' reactions highlighted three key weaknesses concerning U.S. intelligence support:

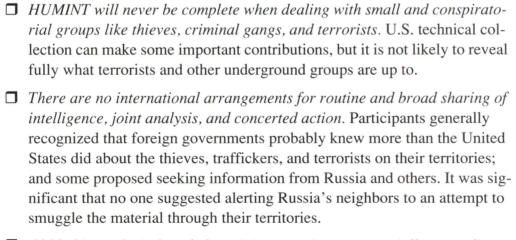

❑ *HUMINT will never be complete when dealing with small and conspiratorial groups like thieves, criminal gangs, and terrorists.* U.S. technical collection can make some important contributions, but it is not likely to reveal fully what terrorists and other underground groups are up to.

❑ *There are no international arrangements for routine and broad sharing of intelligence, joint analysis, and concerted action.* Participants generally recognized that foreign governments probably knew more than the United States did about the thieves, traffickers, and terrorists on their territories; and some proposed seeking information from Russia and others. It was significant that no one suggested alerting Russia's neighbors to an attempt to smuggle the material through their territories.

❑ *Old habits and mindsets led participants to ignore potentially rewarding options.* Participants were slow to suggest approaching Iran once it appeared that a rift had opened between Tehran and Hizballah, and only in the IWG sessions did anyone suggest probing for disaffected members of Hizballah or any Chechens who might cooperate to prevent a nuclear attack.

The arrangements that might help track stolen material, monitor terrorist groups, or intercept nuclear devices as far as possible from U.S. shores go beyond what the United States can do unilaterally or via traditional bilateral liaison. Wild Atom controllers concluded that nuclear smuggling and terrorism require a fresh approach, including multilateral liaison to

❑ pass timely warning of an emerging threat;

❑ pool information on a routine basis;

❑ conduct joint assessments to identify trafficker and terrorist profiles, routes, and methods; and

❑ interlock national defenses so they cannot be bypassed.

Wild Atom also explored technical capabilities for detecting nuclear materials (discussed later in the Technology section of this chapter) and covert action options to neutralize the threat. Although one participant introduced an array of covert action proposals in the Intelligence IWG, nothing was passed during the subsequent NSC session. Covert action (using agents of influence, disinformation and propaganda, and technical detectors improved to 2001 standards that are carried by teams on the ground or aboard remotely piloted vehicles overflying the Bekaa Valley) offers possibilities to buy time, locate nuclear devices, deter their use, entrap terrorists, or orchestrate the actions of foreign governments that seek independently to neutralize the threat.

Organization

U.S. Government

Combating post–Cold War threats like nuclear terrorism requires close cooperation among an expanded array of U.S. agencies, including some that do not normally participate in national security deliberations. In Wild Atom, the prospect that nuclear terrorists were coming to America dictated cooperation between foreign intelligence operatives and JCS-DoD officers on the one hand and domestic law enforcement officers, nuclear-weapon experts, environmental cleanup specialists, and public health experts on the other hand.

Viewing the entities represented around the various tables, participants in Wild Atom opined that institutional arrangements and personal patterns of cooperation in the U.S. government today are ineffective for coping with a crisis along the lines of our scenario. They identified two areas in need of urgent attention:

❑ *Weak linkages within and among the policy, intelligence, law enforcement, and technical communities should be strengthened.* Because U.S. officials persist in operating in separate niches or stovepipes, agencies are better prepared individually to deal with nuclear smuggling and terrorism than they are collectively. Even though Wild Atom compelled participants to work together in one of three communities, participants correctly realized that they needed inputs from the other IWGs.

❑ *A lead agency or single forum should be designated for nuclear smuggling and nuclear terrorism.* Indeed, the greater issue is terrorism and proliferation of all weapons of mass destruction, not merely nuclear weapons. At present, however, participants said that neither the nonproliferation nor the counterterrorism community is in charge. (In the parlance of corporate "reengineers," this is an example of a dysfunctional organization fragmenting a single process. Sequential steps to block terrorist acquisition, development, and use of nuclear capabilities are all one process and should be under the control of a single manager.)

Uncomfortable at working with each other, participants tended to do what they knew how to do instead of what needed doing. CIA and FBI participants pursued their separate intelligence or police agendas. When some participants proposed deploying technical sensors aboard Coast Guard and Navy ships to detect the threat before it reached U.S. shores, others observed that there was no time to train new operators and they doubted that the devices could be used effectively at sea. Not appearing to understand that the game is lost if a device enters a U.S. port (because it could be detonated aboard ship), some participants preferred to retain the inadequate defenses the United States had rather than devise new ones better suited to the situation.

Representatives from FEMA and the Public Health Service observed that they rarely attended meetings or worked routinely with others on the national security team. Moreover, despite improving cooperation between the departments of energy and defense, participants said domestic planning lags behind what the United States is preparing to do overseas. Civil preparedness is improving, but continuity of government planning seems stuck in a Cold War mode and—despite the continuing maturity of the National Disaster Medical System (NDMS) and other initiatives—integration of domestic state, regional, and local emergency, health, and other service providers is lethargic and intermittent.

All agreed that preparations for nuclear consequence management must start well before the incident but that this is not yet happening sufficiently. For example, senior U.S. officials should prerecord announcements advising the public to stay calm and at home—or, alternatively, to evacuate in a prompt but orderly fashion—so that the media could immediately air authoritative and appropriate instructions in a crisis, but such recordings do not exist.

And some participants, despite long years of government service, were simply uncomfortable suggesting concrete actions to be taken. Intelligence analysts long indoctrinated not to prescribe policy or not to risk having their objective analysis politicized by overly close relations with policymakers resisted crafting policy options.

International

Participants said the exercise highlighted that arrangements for international sharing of intelligence and other resources in a crisis must be in place and exercised frequently if they are to function effectively when needed. Such international arrangements do not yet exist, and current efforts to obtain them are inadequate. The Law Enforcement IWG was pessimistic about gaining greater international cooperation. New forums, agreements, confidence-building measures, and joint exercises are all needed, they said, to overcome each state's natural inclination to go it alone and concentrate on keeping nuclear devices and materials out of its national territory.

Others reinforced the pessimistic outlook by pointing to specific obstacles to closer international collaboration against nuclear terrorists. Rules that hinder the sharing of classified or nuclear-energy-controlled information, differences in training and equipment, and a residue of Cold War distrust are among the barriers to joint operations with Russia, despite confident predictions to the contrary by U.S. officials. The Science/Technology IWG insisted that limited U.S. consequence-management resources should not be shared abroad even though, according to the scenario, the U.S. president had already offered full assistance to the Russian government in dealing with the nuclear detonation outside Moscow.

Policy

Deliberations in the simulated Principals Committee and NSC meetings were high level, charged, and very much concerned with what to tell the American people. But they were insufficiently operational, in part because few of the principal officers presented action proposals developed in their IWG sessions.

The exercise highlighted the lack of a national policy to deal with a nuclear terrorist crisis.

One frustrated and very senior participant suggested that the United States should publicly announce a tough policy toward nuclear terrorists that controllers summarized as "you will die if you try." In the Wild Atom context, the participant suggested three steps:

❑ *Defense.* Close U.S. borders for 48 hours and push the nuclear detection effort offshore.

❑ *Deterrence.* Declare that any entity possessing nonsafeguarded nuclear material must give it up or be considered fair game for U.S. preemptive action. The United States should also increase rewards for surrendering fissile material or providing information on undeclared stocks or nuclear traffickers.

❑ *Response.* Eliminate any entity that causes a nuclear explosion. Brand nuclear criminals as international outlaws, and invoke a firm policy of relentless pursuit of nuclear terrorists, much as law enforcement agencies today chase down "cop killers."

The U.S. government should carefully determine—in unhurried deliberation before the heat of a crisis—whether to negotiate with nuclear terrorists. Although this was the consensus reached, the participants were divided on what that policy should be. One rejected negotiation because negotiating would encourage more attempts at nuclear extortion. Another approved because a nuclear disaster must be avoided, and any payment could be finessed as a reward.

Participants identified other crucial questions that should be decided in advance of a possible crisis: Given the danger and liability, should the United States help recover or render safe a nuclear device abroad? Under what conditions should the government evacuate a U.S. city?

The issue that occupied a full one-third of both the Principals Committee and NSC meetings was what to tell the American people about the crisis. The debate was between tough talk designed to deter would-be nuclear terrorists and reassurances to avoid public panic; for example:

> **The President:** Now, Mr. Director [of FEMA], I don't want the press six months from now saying that we applied a Chernobyl solution, as Mr. Gorbachev did with regard to holding populations in place around Chernobyl. If it is necessary for us to move people out, we should do so.

Over the longer term, there was support for a U.S. information policy to deter nuclear crimes. Such a policy should spread the word that nuclear traffickers are risking their health and lives, are unlikely to conclude a profitable sale, and are most likely to be caught and punished severely.

Technology

> The most credible threat [Hizballah] poses to the continental United States is a radioactive dispersal device using the plutonium and a high-explosive device to disperse it in a large area.
>
> However, if Hizballah is assumed to have technical assistance from the Iranians or others, then our assessment changes. And, in fact, they have the capability to fabricate a plutonium implosion device; three devices, in fact.
>
> —Secretary of Energy

The varying technological capabilities of the Iranian government and Hizballah prompted the secretary of energy to declare that the most critical intelligence requirement was determining exactly who had the nuclear material and what they intended to do with it. Most others, however, focused on locating the material and intercepting whatever device had been made of it. Detecting the material posed a major technological challenge.

There was broad consensus that the United States needed to develop new, inexpensive, compact, stand-off, multiple-threat detectors. Most participants doubted, however, that current research funding and development efforts would meet that goal in a timely fashion.

Nonetheless, participants judged that technical detection is a valuable—but possibly overvalued—tool in combating nuclear terrorism. Given the many possible targets, approach routes, and means of transport, technical detectors will always be too few. Moreover, some fissile materials are difficult to detect, and shielding (or distance) can defeat even that limited capability. Existing detectors can be useful in the hands of facility guards, customs inspectors, police, and intelligence operatives.

Reflections

Wild Atom provided a structured environment for thinking through and discussing issues related to nuclear terrorism. Most participants gained a greater awareness of the potential threat and valuable insights into countermeasures. They also had an opportunity to experience stressful decision-making, glean lessons from the experience, and contemplate what needs to be put in place well before a possible nuclear terrorism crisis.

The exercise had a number of important limitations, however. It was all too brief because the participants are busy people. It was interactive only among the players; thus the scenario was not altered to force participants to confront the consequences of their decisions. It avoided injecting new information while meetings were in progress in order to focus on in-depth discussion of core issues and not push the players to respond quickly to a series of stimuli. Regarding the content of the exercise, were the Nuclear Black Market Task Force to conduct another exercise, the task force would explore further certain dimensions of the problem: organized criminal involvement, the brain drain from the former Soviet weapon states as a source of nuclear expertise, potential use of the Internet and fax technology for transmitting weapon designs and targeting data, and consequence management of a nuclear detonation in a U.S. city.

The approach used in Wild Atom can be adapted to domestic or international training related to nuclear smuggling and nuclear terrorism or broadened to include terrorism that makes use of various weapons of mass destruction. We hope that others will use Wild Atom to raise awareness of this potential threat and to help responsible officials prepare to deal with it.

Wild Atom Participants

Nicholas Asimakopoulos
CSIS

Michael Austin
*Federal Emergency Management
    Agency*

Michael Bopp
*Senate Permanent Subcommittee on
    Investigations*

Kevin Callahan
Nonproliferation Center

Burrus M. Carnahan
*Science Applications International
    Corporation (SAIC)*

Frank Cilluffo
CSIS

Craig Conklin
U.S. Environmental Protection Agency

Guild Copeland
CSIS

Anthony H. Cordesman
CSIS

Carla Corliss
CSIS

John Davidson
U.S. Nuclear Regulatory Commission

Arnaud de Borchgrave
CSIS

Paul Dembnicki
Federal Bureau of Investigation

John Dillard
Nonproliferation Center

Mark Dowd
Federal Bureau of Investigation

John J. Dziak
Dziak Group, Inc.

Robert E. Ebel
CSIS

Ray Firehock
*Arms Control and Disarmament
    Agency*

James Ford
National Defense University

Frank Frysiek
U.S. Customs Service

Richard Galbraith
U.S. Customs Service

Harry W. Geiglein
U.S. Secret Service

Curt Gergely
Booz-Allen & Hamilton Inc.

Lyndel R. Hardy
U.S. Secret Service

DeCassandra Harris
CSIS

Carl Henry
*Los Alamos National Laboratory
    (NEST Team)*

John D. Holum
*Arms Control and Disarmament
    Agency*

Fred C. Iklé
Former Under Secretary of Defense

John D. Immele
U.S. Department of Energy

Lieutenant Colonel John Kierepka
(USMC)
Defense Special Weapons Agency

Erik Kjonnerod
Institute for National Strategic Studies

Nicholas Koukopoulos
CSIS

Robert H. Kupperman
CSIS

Rear Admiral John L. Linnon (USCG)
U.S. Department of Transportation

Gregory Lyle
Defense Special Weapons Agency

Douglas McEachin
*Former Deputy Director of
    Intelligence
Central Intelligence Agency*

General Edward C. Meyer
U.S. Army (Ret.)

David Miller
CSIS

Steve Mladineo
*Battelle Pacific Northwest National
    Laboratory*

Sarah A. Mullen
*Arms Control and Disarmament
    Agency*

Bruce Murray
U.S. Customs Service

Jon Neasham
Defense Special Weapons Agency

Brian Nordmann
*Arms Control and Disarmament
    Agency*

Richard Norris
Defense Intelligence Agency

Lawrence O'Donnell
U.S. Customs Service

Erik R. Peterson
CSIS

Rear Admiral Paul J. Pluta (USCG)
U.S. Department of Transportation

Renee Pruneau
Nonproliferation Center

Linnea Raine
CSIS

Victoria Reznik
Institute for National Strategic Studies

James R. Schlesinger
*Former Secretary of Defense
Former Secretary of Energy
Former Director, Central Intelligence
    Agency*

Barbara Shaw
Defense Intelligence Agency

Jeff Shumaker
U.S. Department of Transportation

Michael Smith
U.S. Department of Transportation

James Stekert
U.S. Army

Jessica Stern
Former National Security Council Staff

John B. Stewart
*Ogden Environmental and Energy
    Services Inc.*

Kenneth Stroech
U.S. Environmental Protection Agency

George J. Terwilliger
Former Deputy Attorney General

J. Stephen Veyera
Federal Bureau of Investigation

Daniel Wagner
Central Intelligence Agency

Colonel Katherine Ward (USA)
Institute for National Strategic Studies

William H. Webster
*Former Director, Central Intelligence
    Agency
Former Director, Federal Bureau of
    Investigation*

Berthold W. Weinstein
*Lawrence Livermore National
    Laboratory*

Commander Peter Wikul (USN)
Joint Chiefs of Staff

Admiral Webster Young
U.S. Public Health Service

Chronology

1995	Hizballah sets up facility in Lebanon's Bekaa Valley; begins to acquire wherewithal to manufacture nuclear weapons; soon lacks only design experts and fissile material.
1997	Chechen separatists establish secret facility in Rostov, Russia, to make nuclear arms; begin to acquire tools, materials, and former Soviet weapons experts; eventually lack only fissile material.

2000

12 Dec	Chechen separatists agree to buy 200 kg. of weapons-grade HEU from local Chechen gang in Chelyabinsk. Separately, Iran agrees to buy 20 kg. of weapons-grade plutonium, enough for two HEU and three plutonium weapons of 10–15-kiloton yield.
13 Dec	Hizballah learns of Tehran's plan to buy plutonium and makes plans to steal it upon arrival in Tehran; concludes hiring two former Soviet nuclear-weapons experts; authorizes purchase of nuclear triggers and chemical explosives from South Africa.
14 Dec	Theft from Chelyabinsk.
15 Dec	HEU delivered to Chechen separatists in Rostov. Five firms in Europe and in Asia wire $23.5 million to Zurich bank.
28 Dec	Plutonium arrives in Tehran; Detroit bank wires $20 million to same Zurich bank; Hizballah steals plutonium.
29 Dec	Iran tries to recover plutonium; attacks Hizballah and arrests known Hizballah sympathizers.
30 Dec	Plutonium arrives at Hizballah camp in Lebanon's Bekaa Valley. Hizballah orders agent in Atlanta to surveil port cities in eastern United States.

2001

26 Jan	First Chechen gun-type bomb complete; moved immediately to Kursk, Russia.
27 Jan	Chechen bomb arrives in Kursk. First Hizballah plutonium implosion bomb complete; moved immediately to Greek freighter waiting in Beirut harbor.

| 30 Jan | Freighter departs Beirut for Baltimore. |

Phase One: Ambiguous Warning

| 8 Feb | Second Chechen bomb complete; dispatched to Moscow. CIA source learns Iranians believe Hizballah intends nuclear attacks on the United States and its allies. |

| 9 Feb | U.S. technicians arrive at Chelyabinsk. Russians discover fissile material is missing. Communication triples among Russian leaders and nuclear units. Recovery operations commence. U.S. National Security Agency initiates NOIWON alert; U.S. president calls President Lebed. |

| 10 Feb | Russians confine U.S. technicians. U.S. Principals Committee meets. |

Phase Two: Crisis Management

| 12 Feb | U.S. satellite detects nuclear detonation outside Moscow; President Lebed calls U.S. president; Chechen separatists claim credit. Hizballah issues ultimatum. New York harbor police arrest Iranian student. FBI activates Federal Emergency Response Plan. U.S. NSC meets. |

| 14 Feb | Greek freighter due to arrive in Baltimore harbor. Hizballah plans to detonate device on ship. |